CHEMISTRY ESSENTIALS: ORGANIC AND INORGANIC

Edited by
Jean Annerie Hernandez

Delve Publishing

Chemistry Essentials: Organic and Inorganic

Edited by Jean Annerie Hernandez

Edition 2018

Delve Publishing

Published by
Delve Publishing LLC
708 3rd Avenue, 6th Floor
New York
NY 10801
United States of America
www.delvepublishing.com

ISBN: 978-1-68095-726-6 (Hardback)

Printed in United States of America

Library of Congress Control Number: 2016920406

About the Editor

Jean Annerie Hernandez

Jean Annerie took up a Bachelor's degree in Chemistry in Central Luzon State University before she decided to fulfill her lifelong dream of becoming a creative writer and editor. She has ghostwritten and copyedited fictional and non-fictional books for three years, and is currently working on various client-based projects in Singapore, China and USA.

Table of Contents

3 NOMENCLATURE OF ORGANIC CHEMISTRY

4 THEORY OF ORGANIC REACTIONS

5 ORGANIC SYNTHESIS

6 THEORY OF AROMATICITY

7 CHEMICAL BONDING

8 ACID-BASE CHEMISTRY

9 IONIC AND COVALENT SOLIDS - STRUCTURES

10 REDOX STABILITY AND REDOX REACTIONS

List of Abbreviations

ATP	Adenosine Triphosphate
BE	Binding Energy
CBC	Covalent Bond Classification
CFSE	Crystal Field Stabilization Energy
FASI	Fatty Acid Synthase I
IDL	Intermediate-Density Lipoprotein,
IARC	International Agency for Research on Cancer
IUPAC	International Union of Pure and Applied Chemistry
ISM	Interstellar Medium
LEDs	Light-Emitting Diodes
LCAO	Linear Combinations of Atomic Orbitals
LDL	Low-Density Lipoprotein,
MO	Molecular Orbital,
PAHs	Polycyclic Aromatic Hydrocarbons, , ,
VB	Valence Bond Theory
VSEPR	Valence Shell Electron-Pair Repulsion
VLDL	Very Low-Density Lipoproteins

Preface

Organic chemistry is a highly creative science in which chemists create new molecules and explore the properties of existing compounds. Organic compounds are all around us. They are central to the economic growth in the rubber, plastics, fuel, pharmaceutical, cosmetics, detergent, coatings, dyestuff, and agrichemical industries, to name a few. In organic chemistry, molecules that have similar properties (whether they are chemical or physical properties) are grouped together. The reason they have similar properties is because they have similar groups of atoms; these groups of atoms are called functional groups. Chemical properties involve one substance changing into another substance by reacting. An example of a chemical property is the ability of chlorine gas to react explosively when mixed with sodium. The chemical reaction creates a new substance, sodium chloride. Physical properties refer to different forms of a substance, but the substance remains the same; no chemical reaction or change to a new substance occurs. Some of the properties that the functional groups provide include polarity and acidity. For example, the functional group called carboxyl (-COOH) is a weak acid. Polarity refers to one end of a molecule having a charge (polar), and the other end having no charge (nonpolar). For example, the plasma membrane has hydrophilic heads on the outside that are polar, and the hydrophobic tails (which are nonpolar) form the inside of the plasma membrane. However, Inorganic Chemistry is not an isolated branch of chemistry. This core science is fully unified with other areas of chemistry such as organic, physical and analytical chemistry. It deals with the chemistry of all non-organic compounds, and mainly involves the chemistry of metals and especially transition metals. These elements play a crucial role in industrial catalytic processes that are required to produce substances and new materials at a rate far exceeding that of natural chemical reactions. Such catalytic processes can take place in solution or on the surface of solid materials and usually involve transition metal elements.

Content Coverage

Chapters One and *Two* are intended to present an overview of covalent bonding and functional groups, correspondingly. Covalent bonding includes many kinds of interactions, including σ-bonding, π-bonding, metal-to-metal bonding, agostic interactions, bent bonds, and three-center two-electron bonds.

Chapters Three presents the information about nomenclature system of organic compounds. As organic chemistry grew and developed, many compounds were given trivial names, which are now commonly used and recognized.

Chapters Four and *Five* include facts about the theory of organic reactions and organic synthesis, respectively. In organic synthesis, organic reactions are used in the construction of new organic molecules. The production of many man-made chemicals such as drugs, plastics, food additives, fabrics depend on organic reactions.

Chapter Six focuses on the theory of aromaticity used to describe a cyclic (ring-shaped), planar (flat) molecule with a ring of resonance bonds that exhibits more stability than other geometric or connective arrangements with the same set of atoms.

Chapters Seven and *Eight* are intended to focus on chemical bonding and acid-base reactions, separately. Acid-base reactions are chemical reactions that occur between acids and bases.

Chapter Nine aims to systematize the structures of inorganic solids - metal oxides, halides, sulfides, and related compounds - and develop some rules for which structures to expect based on electronegativity differences, hard-soft acid-base rules, and other periodic trends.

Chapter Ten is aimed at redox stability and redox reactions. In terms of everyday life, redox reactions occur all of the time around us, such as, the metabolism of sugars to CO_2, which stores energy in the form of ATP, is a redox reaction.

CHAPTER 1

Covalent Bond

Chapter Contents

- **Introduction**
- **History**
- **Typesof Covalent Bonds**
- **Covalent Structures**
- **One-and Three-Electron Bonds**
- **Resonance**
- **Quantum Mechanical Description**
- **Covalent Bond Classification Method**
- **Chemical Bonding Model**
- **VS**
- **Models for Transition Metals**

In covalent bonding, the atoms are unstable because their outer rings of electrons aren't filled up. By sharing electrons with other atoms, these atoms can fill up their outer rings and become stable. In water, for instance, the oxygen atom needs two more electrons to be stable, and the hydrogen atoms each need one. When they get together, the oxygen atom shares one electron with each of the hydrogen atoms, and the hydrogen atoms each share one electron with the oxygen atom. Covalent bonds are usually weaker than ionic bonds.

INTRODUCTION

A **covalent bond**, also called a **molecular bond**, is a chemical bond that involves the sharing of electron pairs between atoms. These electron pairs are known as **shared pairs** or **bonding pairs**, and the stable balance of attractive and repulsive forces between atoms, when they share electrons, is known as covalent bonding. For many molecules, the sharing of electrons allows each atom to attain the equivalent of a full outer shell, corresponding to a stable electronic configuration.

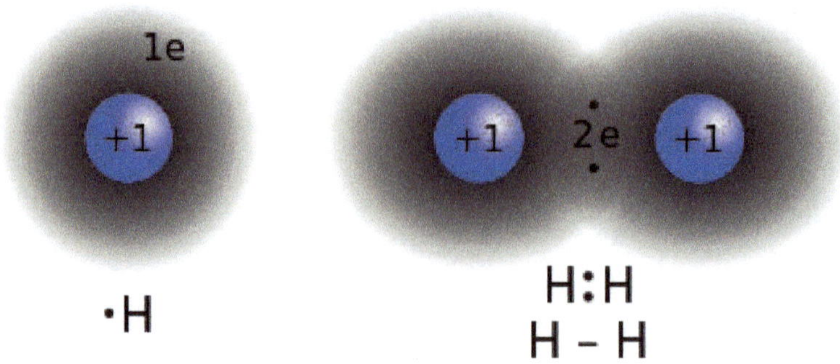

A covalent bond forming H_2 (right) where two hydrogen atoms share the two electrons.

Covalent bonding includes many kinds of interactions, including σ-bonding, π-bonding, metal-to-metal bonding, agostic interactions, bent bonds, and three-center two-electron bonds. The term *covalent bond* dates from 1939. The prefix *co-* means *jointly, associated in action, partnered to a lesser degree,* etc.; thus a "co-valent bond", in essence, means that the atoms share "valence , such as is discussed in valence bond theory.

In the molecule H_2, the hydrogen atoms share the two electrons via covalent bonding. Covalency is greatest between atoms of similar electronegativities. Thus, covalent bonding does not necessarily require that the two atoms be of the same elements, only that they be of comparable electronegativity. Covalent bonding that entails sharing of electrons over more than two atoms is said to be delocalized.

HISTORY

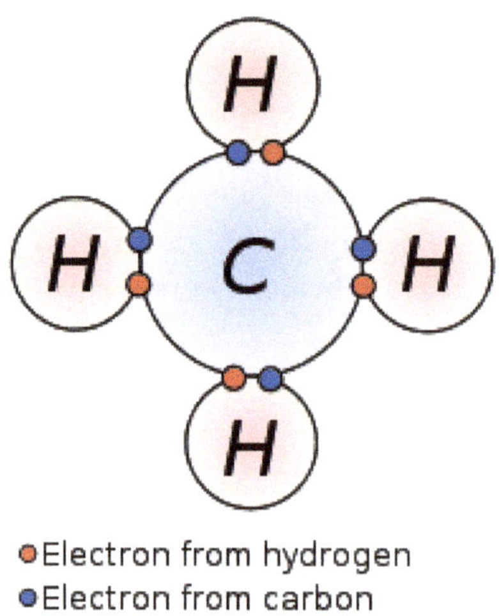

Early concepts in covalent bonding arose from this kind of image of the molecule of methane. Covalent bonding is implied in the Lewis structure by indicating electrons shared between atoms.

The term *covalence* in regard to bonding was first used in 1919 by Irving Langmuir in a *Journal of the American Chemical Society* article entitled "The Arrangement of Electrons in Atoms and Molecules". Langmuir wrote that "we shall denote by the term *covalence* the number of pairs of electrons that a given atom shares with its neighbors."

The idea of covalent bonding can be traced several years before 1919 to Gilbert N. Lewis, who in 1916 described the sharing of electron pairs between atoms. He introduced the *Lewis notation* or *electron dot notation* or *Lewis dot structure*, in which valence electrons (those in the outer shell) are represented as dots around the atomic symbols. Pairs of electrons located between atoms represent covalent bonds. Multiple pairs represent multiple bonds, such as double bonds and triple bonds. An alternative form of representation, not shown here, has bond-forming electron pairs represented as solid lines.

Lewis proposed that an atom forms enough covalent bonds to form a full (or closed) outer electron shell. In the methane diagram shown here, the carbon atom has a valence of four and is, therefore, surrounded by eight electrons (the octet rule), four from the carbon itself and four from the hydrogens bonded to it. Each hydrogen has a valence of one and is surrounded by two electrons (a duet rule) – its own one electron plus one from the carbon. The numbers of electrons correspond to full shells in the quantum theory of the atom; the outer shell of a carbon atom is the $n = 2$ shell, which can hold eight electrons, whereas the outer (and only) shell of a hydrogen atom is the $n = 1$ shell, which can hold only two.

While the idea of shared electron pairs provides an effective qualitative picture of covalent bonding, quantum mechanics is needed to understand the nature of these bonds and predict the structures and properties of simple molecules. Walter Heitler and Fritz London are credited with the first successful quantum mechanical explanation of a chemical bond (molecular hydrogen) in 1927. Their work was based on the valence bond model, which assumes that a chemical bond is formed when there is good overlap between the atomic orbitals of participating atoms.

TYPES OF COVALENT BONDS

Atomic orbitals (except for s orbitals) have specific directional properties leading to different types of covalent bonds. Sigma (σ) bonds are the strongest covalent bonds and are due to head-on overlapping of orbitals on two different atoms. A single bond is usually a σ bond. Pi (π) bonds are weaker and are due to lateral overlap between p (or d) orbitals. A double bond between two given atoms consists of one σ and one π bond, and a triple bond is one σ and two π bonds.

Covalent bonds are also affected by the electronegativity of the connected atoms which determines the chemical polarity of the bond. Two atoms with equal electronegativity will make nonpolar covalent bonds such as H–H. An unequal relationship creates a polar covalent bond such as with H−Cl.

COVALENT STRUCTURES

There are several types of structures for covalent substances, including individual molecules, molecular structures, macromolecular structures and giant covalent structures. Individual molecules have strong bonds that hold the atoms together, but there are negligible forces of attraction between molecules. Such covalent substances are usually gases, for example, HCl, SO_2, CO_2, and CH_4. In molecular structures, there are weak forces of attraction. Such covalent substances are low-boiling-temperature liquids (such as ethanol), and low-melting-temperature solids (such as iodine and solid CO_2). Macromolecular structures have large numbers of atoms linked by covalent bonds in chains, including synthetic polymers such as polyethylene and nylon, and biopolymers such as proteins and starch. Network covalent structures (or giant covalent structures) contain large numbers of atoms linked in sheets (such as graphite), or 3-dimensional structures (such as diamond and quartz). These substances have high melting and boiling points, are frequently brittle, and tend to have high electrical resistivity. Elements that have high electronegativity, and the ability to form three or four electron pair bonds, often form such large macromolecular structures.

ONE- AND THREE-ELECTRON BONDS

Bonds with one or three electrons can be found in radical species, which have an odd number of electrons. The simplest example of a 1-electron bond is found in the dihydrogen cation, H+2. One-electron bonds often have about half the bond energy of a 2-electron bond, and are therefore called half bonds . However, there are exceptions: in the case of dilithium, the bond is actually stronger for the 1-electron Li+2 than for the 2-electron Li_2. This exception can be explained in terms of hybridization and inner-shell effects.

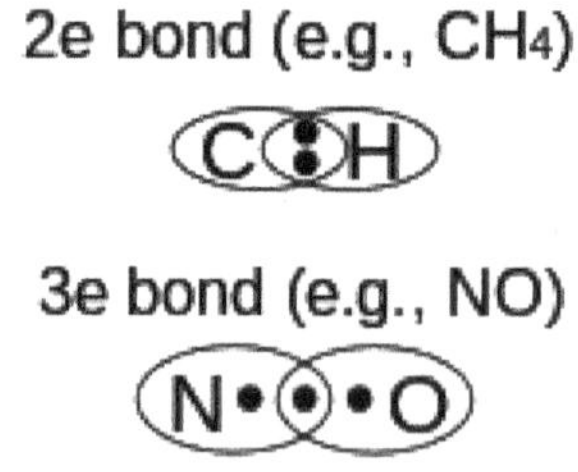

Comparison of the electronic structure of the three-electron bond to the conventional covalent bond.

The simplest example of three-electron bonding can be found in the helium dimer cation, He+2. It is considered a “half bond” because it consists of only one shared electron (rather than two); in molecular orbital terms, the third electron is in an anti-bonding orbital which cancels out half of the bond formed by the other two electrons. Another example of a molecule containing a 3-electron bond, in addition to two 2-electron bonds, is nitric oxide, NO. The oxygen molecule, O_2 can also be regarded as having two 3-electron bonds and one 2-electron bond, which accounts for its paramagnetism and its formal bond order of 2. Chlorine dioxide and its heavier analogues bromine dioxide and iodine dioxide also contain three-electron bonds.

Molecules with odd-electron bonds are usually highly reactive. These types of bond are only stable between atoms with similar electronegativities.

RESONANCE

There are situations whereby a single Lewis structure is insufficient to explain the electron configuration in a molecule, hence a superposition of structures are needed. The same two atoms in such molecules can be bonded differently in different structures (a single bond in one, a double bond in another, or even none at all), resulting in a non-integer bond order. The nitrate ion is one such example with three equivalent structures. The bond between the nitrogen and each oxygen is a double bond in one structure and a single bond in the other two, so that the average bond order for each N–O interaction is $\frac{2+1+1}{3} = \frac{4}{3}$.

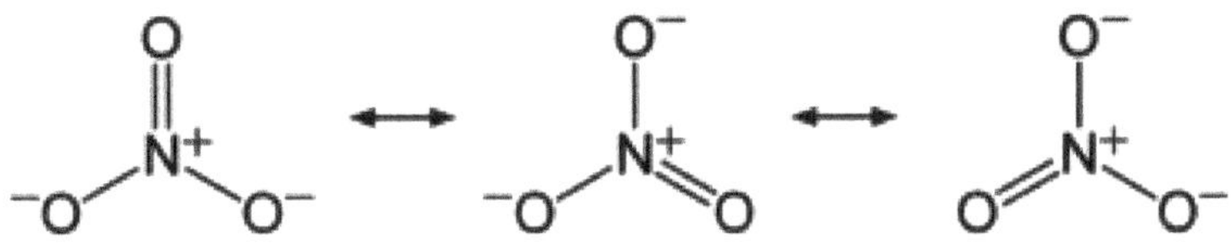

Aromaticity

In organic chemistry, when a molecule with a planar ring obeys Hückel's rule, where the number of π electrons fit the formula $4n + 2$ (where n is an integer), it attains extra stability and symmetry. In benzene, the prototypical aromatic compound, there are 6 π bonding electrons ($n = 1$, $4n + 2 = 6$). These occupy three delocalized π molecular orbitals (molecular orbital theory) or form conjugate π bonds in two resonance structures that linearly combine (valence bond theory), creating a regular hexagon exhibiting a greater stabilization than the hypothetical 1,3,5-cyclohexatriene.

In the case of heterocyclic aromatics and substituted benzenes, the electronegativity differences between different parts of the ring may dominate the chemical behaviour of aromatic ring bonds, which otherwise are equivalent.

Hypervalence

Certain molecules such as xenon difluoride and sulfur hexafluoride have higher co-ordination numbers than would be possible due to strictly covalent bonding according to the octet rule. This is explained by the three-center four-electron bond ("3c–4e") model which interprets the molecular wavefunction in terms of non-bonding highest occupied molecular orbitals in molecular orbital theory and ionic-covalent resonance in valence bond theory.

Electron-Deficiency

In three-center two-electron bonds ("3c–2e") three atoms share two electrons in bonding. This type of bonding occurs in electron deficient compounds like diborane. Each such bond (2 per molecule in diborane) contains a pair of electrons which connect the boron atoms to each other in a banana shape, with a proton (nucleus of a hydrogen atom) in the middle of the bond, sharing electrons with both boron atoms. In certain cluster compounds, so-called four-center two-electron bonds also have been postulated.

QUANTUM MECHANICAL DESCRIPTION

After the development of quantum mechanics, two basic theories were proposed to provide a quantum description of chemical bonding: valence bond (VB) theory and molecular orbital (MO) theory. A more recent quantum description is given in terms of atomic contributions to the electronic density of states.

Covalency from Atomic Contribution to the Electronic Density of States

In COOP, COHP and BCOOP, evaluation of bond covalency is dependent on the basis set. To overcome this issue, an alternative formulation of the bond covalency can be provided in this way.

The center mass $cm(n,l,m_l,m_s)$ of an atomic orbital $|n,l,m_l,m_s\rangle$, with quantum numbers n, l, m_l, m_s, for atom A is defined as

$$cm^{\mathrm{A}}(n,l,m_l,m_s) = \frac{\int_{E_0}^{E_1} E g^{\mathrm{A}}_{|n,l,m_l,m_s\rangle}(E)\,dE}{\int_{E_0}^{E_1} g^{\mathrm{A}}_{|n,l,m_l,m_s\rangle}(E)\,dE}$$

where $g^{\mathrm{A}}_{|n,l,m_l,m_s\rangle}(E)$ is the contribution of the atomic orbital $|n,l,m_l,m_s\rangle$ of the atom A to the total electronic density of states $g(E)$ of the solid

$$g(E) = \sum_{\mathrm{A}} \sum_{n,l} \sum_{m_l,m_s} g^{\mathrm{A}}_{|n,l,m_l,m_s\rangle}(E)$$

where the outer sum runs over all atoms A of the unit cell. The energy window $[E_0,E_1]$ is chosen in such a way that it encompasses all relevant bands participating in the bond. If the range to select is unclear, it can be identified in practice by examining the molecular orbitals that describe the electron density along the considered bond.

The relative position $C_{n\mathrm{A}l\mathrm{A},n\mathrm{B}l\mathrm{B}}$ of the center mass of $|n_{\mathrm{A}},l_{\mathrm{A}}\rangle$ levels of atom A with respect to the center mass of $|n_{\mathrm{B}},l_{\mathrm{B}}\rangle$ levels of atom B is given as

$$C_{n_{\mathrm{A}}l_{\mathrm{A}},n_{\mathrm{B}}l_{\mathrm{B}}} = -\left|cm^{\mathrm{A}}(n_{\mathrm{A}},l_{\mathrm{A}}) - cm^{\mathrm{B}}(n_{\mathrm{B}},l_{\mathrm{B}})\right|$$

where the contributions of the magnetic and spin quantum numbers are summed. According to this definition, the relative position of the A levels with respect to the B levels is

$$C_{\mathrm{A,B}} = -\left|cm^{\mathrm{A}} - cm^{\mathrm{B}}\right|$$

where, for simplicity, we may omit the dependence from the principal quantum number n in the notation referring to $C_{n\mathrm{A}l\mathrm{A},n\mathrm{B}l\mathrm{B}}$.

In this formalism, the greater the value of $C_{\mathrm{A,B}}$, the higher the overlap of the selected atomic bands, and thus the electron density described by those orbitals gives a more covalent A–B bond. The quantity $C_{\mathrm{A,B}}$ is denoted as the *covalency* of the A–B bond, which is specified in the same units of the energy E.

COVALENT BOND CLASSIFICATION METHOD

The covalent bond classification (CBC) **method** is also referred to as the LXZ notation. It was published by M. L. H. Green in the mid-1990s as a solution for the need to describe covalent compounds such as organometallic complexes in a way that is not prone to limitations resulting from the definition of oxidation state. Instead of simply assigning a charge to an atom in the molecule (i.e. the oxidation state), the covalent bond classification method analyzes the nature of the ligands surrounding the atom of interest, which is often a transition metal. According to this method, there are three basic types of interactions that allow for coordination of the ligand. The three types of interaction are classified according to whether the ligating group donates two, one, or zero electrons. These three classes of ligands are respectively given the symbols L, X and Z.

Types of Ligands

X-type ligands are those that donate one electron to the metal and accept one electron from the metal when using the neutral ligand method of electron counting, or donate two electrons to the metal when using the donor pair method of electron counting. Regardless of whether it is considered neutral or anionic, these ligands yield normal covalent bonds. A few examples of this type of ligand are H, halogens (Cl, Br, F, etc.), OH, CN, CH_3, and NO (bent).

L-type ligands are neutral ligands that donate two electrons to the metal center regardless of the electron counting method being used. These electrons can come from lone pairs, pi or sigma donors. The bonds formed between these ligands and the metal are dative covalent bonds, which are also known as coordinate bonds. Examples of this type of ligand include CO, PR_3, NH_3, H_2O, carbenes (=CRR'), and alkenes.

Z-type ligands are those that accept two electrons from the metal center as opposed to the donation occurring with the other two types of ligands. However, these ligands also form dative covalent bonds like the L-type. This type of ligand is not usually used, because in certain situations it can be written in terms of L and X. For example, if a Z ligand is accompanied by an L type, it can be written as X_2. Examples of these ligands are Lewis acids, such as BR_3.

Uses of the Notation

When given a metal complex and the trends for the ligand types, the complex can be written in a more simplified manner with the form $[ML_lX_xZ_z]^{Q\pm}$. The subscripts represent the numbers of each ligand type present in that complex, M is the metal center and Q is the overall charge on the complex. Some examples of this overall notation are as follows:

condensed formula	LXZ notation
$[Mn(CO)_6]$+	$[ML_6]^+$
$[Ir(CO)(PPh_3)_2(Cl)(NO)]^{2+}$	$[ML_3X_2]^{2+}$

$[Fe(CO)_2(CN)_4]^{2-}$	$[ML_2X_4]^{2-}$

Also from this general form, the values for electron count, oxidation state, coordination number, number of d-electrons, valence number and the ligand bond number can be calculated.

Electron Count = N + x + 2l – Q

Where N is the group number of the metal.

Oxidation State (OS) = x + Q

Coordination Number (CN) = x + l

Number of d-electrons (dn) = N- OS

= N- (x + Q)

Valence Number (VN)= x + 2z

Ligand Bond Number (LBN) = l + x + z

Other Uses

This template of writing a metal complex also allows for better comparison of molecules with different charges. This can happen when the assignment is reduced to its "equivalent neutral class". The equivalent neutral class is the classification of the complex if the charge was localized on the ligand as opposed to the metal center. In other words, the equivalent neutral class is the representation of the complex as though there was no charge.

CHEMICAL BONDING MODEL

Chemical bonding models are theoretical models used to explain atomic bonding structure, molecular geometry, properties, and reactivity of physical matter.

There are a variety of known chemical bonding interactions including covalent, ionic, and metallic bonding among others. The theories associated with bonding are often developed around the covalent bonds and extended to ionic and metallic bonding. There are a variety of active theories or models associated with covalent bonding, the building block of molecules. These theories make various approximations rendering each them useful for describing different nuances of common molecular bonding.

VSEPR

The simplest and most primitive of the theories currently taught. Describes molecular geometry through the repulsion of electron fields which include bonds and lone pairs. It does not require any application of orbital shape.

Wave Function Models

Valence Bond Theory

In 1927, valence bond theory was formulated and it argues that a covalent bond forms when two valence electrons, in their respective atomic orbitals, work or function to hold two nuclei together, by virtue of effects of lowering system energies. Building on this theory, the chemist Linus Pauling published in 1931 what some consider one of the most important papers in the history of chemistry: "On the Nature of the Chemical Bond". In this paper, elaborating on the works of Lewis, and the valence bond theory (VB) of Heitler and London, and his own earlier works, Pauling presented six rules for the shared electron bond, the first three of which were already generally known:

- The electron-pair bond forms through the interaction of an unpaired electron on each of two atoms.
- The spins of the electrons have to be opposed.
- Once paired, the two electrons cannot take part in additional bonds.

His last three rules were new:

- The electron-exchange terms for the bond involve only one wave function from each atom.
- The available electrons in the lowest energy level form the strongest bonds.
- Of two orbitals in an atom, the one that can overlap the most with an orbital from another atom will form the strongest bond, and this bond will tend to lie in the direction of the concentrated orbital.

Building on this article, Pauling›s 1939 textbook: *On the Nature of the Chemical Bond* would become what some have called the bible of modern chemistry. This book helped experimental chemists to understand the impact of quantum theory on chemistry. However, the later edition in 1959 failed to adequately address the problems that appeared to be better understood by molecular orbital theory. The impact of valence theory declined during the 1960s and 1970s as molecular orbital theory grew in usefulness as it was implemented in large digital computer programs. Since the 1980s, the more difficult problems, of implementing valence bond theory into computer programs, have been solved largely, and valence bond theory has seen a resurgence.

Molecular Orbital Theory

Molecular orbitals were first introduced by Friedrich Hund and Robert S. Mulliken in 1927 and 1928. The linear combination of atomic orbitals or "LCAO" approximation for molecular orbitals was introduced in 1929 by Sir John Lennard-Jones. Linear combinations of atomic orbitals (LCAO) can be used to estimate the molecular orbitals that are formed upon bonding between the molecule›s constituent atoms. Similar to an atomic orbital, a Schrödinger equation, which describes the behavior of an electron, can be constructed for a molecular orbital as well. Linear combinations of atomic orbitals, or the sums and differences of the atomic wavefunctions, provide approximate solutions to the Hartree–Fock equations which correspond to the independent-particle approximation of the molecular Schrödinger equation.

When atomic orbitals interact, the resulting molecular orbital can be of three types: bonding, antibonding, or nonbonding.

Bonding MOs:

- Bonding interactions between atomic orbitals are constructive (in-phase) interactions.
- Bonding MOs are lower in energy than the atomic orbitals that combine to produce them.

Antibonding MOs:

- Antibonding interactions between atomic orbitals are destructive (out-of-phase) interactions, with a nodal plane where the wavefunction of the antibonding orbital is zero between the two interacting atoms
- Antibonding MOs are higher in energy than the atomic orbitals that combine to produce them.

Nonbonding MOs:

- Nonbonding MOs are the result of no interaction between atomic orbitals because of lack of compatible symmetries.
- Nonbonding MOs will have the same energy as the atomic orbitals of one of the atoms in the molecule.

Comparison

The two theories differ in the order that the electron configuration of the molecule is built up. For valence bond theory, the atomic hybrid orbitals are filled first to produce a full valence configuration of bonding pairs and lone pairs. If several such configurations exist, a weighted superposition of these configurations is then applied. In contrast, for molecular orbital theory a weighted superposition of atomic orbitals is performed first, followed by the filling of the resulting molecular orbitals by the aufbau principle.

Either theory has its advantages and uses. As valence bond theory builds the molecular wavefunction out of localized bonds, it is more suited for the calculation of bond energies and the understanding of reaction mechanisms. In particular, valence bond theory correctly predicts the dissociation of homonuclear diatomic molecules into separate atoms, while simple molecular orbital theory predicts dissociation into a mixture of atoms and ions. Molecular orbital theory, with delocalized orbitals that obey its symmetry, is more suited for the calculation of ionization energies and the understanding of spectral absorption bands. Molecular orbitals are orthogonal, which significantly increases feasibility and speed of computer calculations compared to nonorthogonal valence bond orbitals.

Although the wavefunctions generated by both theories do not agree and do not match the stabilization energy by experiment, they can be corrected by configuration interaction. This is done by combining the valence bond covalent function with the functions describing all possible ionic configurations or by combining the molecular orbital ground state function with the functions describing all possible excited states using unoccupied orbitals. It can then be seen that the simple molecular orbital approach gives too much weight to the ionic structures while the simple valence bond approach gives too little. This can also be described as saying that the molecular orbital approach neglects electron correlation while the valence bond approach overestimates it.

The two approaches are now regarded as complementary, each providing its own insights into the problem of chemical bonding. Modern calculations in quantum chemistry usually start from (but ultimately go far beyond) a molecular orbital rather than a valence bond approach, not because of any intrinsic superiority in the former but rather because the MO approach is more readily adapted to numerical computations. However, better valence bond programs are now available.

MODELS FOR TRANSITION METALS

Crystal Field Theory

This approximation begins with the geometries of the d orbitals derived from quantum mechanics. Ligands with their electron density are assumed to destabilize the metal d orbitals they interact with raising their energy while the remain d-orbitals drop in energy to balance the overall change in energy.

Ligand Field Theory

Considered a hybrid of CFT and MO Theory or simply an approximate application of MO Theory to transition metal complexes.

Computational Chemistry

Modern computational chemistry applies components of bonding models to simulate various chemical phenomenon associated with bonding. Computational chemistry also extends beyond covalent bonding to investigate the interactions of groups of molecules and higher order structure. In these systems the chemical bond is often simplified and approximated to reduce computing time.

REFERENCES

1. Campbell, Neil A.; Williamson, Brad; Heyden, Robin J. (2006). Biology: Exploring Life. Boston, MA: Pearson Prentice Hall. ISBN 0-13-250882-6. Retrieved 2012-02-05.

2. March, Jerry (1992). Advanced Organic Chemistry: Reactions, Mechanisms, and Structure. John Wiley & Sons. ISBN 0-471-60180-2.
3. Gary L. Miessler; Donald Arthur Tarr (2004). Inorganic Chemistry. Prentice Hall. ISBN 0-13-035471-6.
4. Merriam-Webster – Collegiate Dictionary (2000).
5. Chemical Bonds . Hyperphysics.phy-astr.gsu.edu. Retrieved 2013-06-09.
6. Langmuir, Irving (1919-06-01). The Arrangement of Electrons in Atoms and Molecules . Journal of the American Chemical Society. 41 *(6): 868–934. doi:10.1021/ja02227a002.*
7. Lewis, Gilbert N. (1916-04-01). The atom and the molecule . Journal of the American Chemical Society. 38 *(4): 762–785. doi:10.1021/ja02261a002.*
8. Heitler, W.; London, F. (1927). Wechselwirkung neutraler Atome und homöopolare Bindung nach der Quantenmechanik [Interaction of neutral atoms and homeopolar bonds according to quantum mechanics]. Zeitschrift für Physik. 44 *(6–7): 455–472. doi:10.1007/bf01397394.* English translation in Hettema, H. (2000). Quantum Chemistry: Classic Scientific Papers. World Scientific. p. 140. ISBN 978-981-02-2771-5. Retrieved 2012-02-05.
9. Stranks, D. R.; Heffernan, M. L.; Lee Dow, K. C.; McTigue, P. T.; Withers, G. R. A. (1970). Chemistry: A structural view. Carlton, Vic.: Melbourne University Press. p. 184. ISBN 0-522-83988-6.
10. Weinhold, F.; Landis, C. (2005). Valency and Bonding. Cambridge. pp. 96–100. ISBN 0-521-83128-8.
11. Harcourt, Richard D., ed. (2015). Chapter 2: Pauling 3-Electron Bonds , 4-Electron 3-Centre Bonding, and the Need for an Increased-Valence Theory . Bonding in Electron-Rich Molecules: Qualitative Valence-Bond Approach via Increased-Valence Structures. Springer. ISBN 9783319166766.
12. Pauling, L. (1960). The Nature of the Chemical Bond. Cornell University Press. pp. 340–354.
13. Cammarata, Antonio; Rondinelli, James M. (21 September 2014). Covalent dependence of octahedral rotations in orthorhombic perovskite oxides . Journal of Chemical Physics. 141 *(11): 114704. doi:10.1063/1.4895967. PMID 25240365.*
14. Hughbanks, Timothy; Hoffmann, Roald (2002-05-01). Chains of trans-edge-sharing molybdenum octahedra: metal-metal bonding in extended systems . Journal of the American Chemical Society. 105 *(11): 3528–3537. doi:10.1021/ja00349a027.*
15. Dronskowski, Richard; Bloechl, Peter E. (2002-05-01). Crystal orbital Hamilton populations (COHP): energy-resolved visualization of chemical bonding in solids based on density-functional calculations . The Journal of Physical Chemistry. 97 *(33): 8617–8624. doi:10.1021/j100135a014.*
16. Grechnev, Alexei; Ahuja, Rajeev; Eriksson, Olle (2003-01-01). Balanced crystal orbital overlap population—a tool for analysing chemical bonds in solids . Journal

of Physics: Condensed Matter. 15 *(45): 7751. doi:10.1088/0953-8984/15/45/014. ISSN 0953-8984.*

CHAPTER 2

Functional Groups

Chapter Contents

Pure hydrocarbons, made up of just carbon and hydrogen atoms, make wonderful combustion fuels, such as propane, butane, and the bulk of commercial gasoline. But can you build a living organism out of just combustion fuels? Probably not. Most of the large biological molecules (macromolecules) that make up humans and other organisms are not pure hydrocarbons, but instead contain additional atoms that provide important types of functionality.

INTRODUCTION

In organic chemistry, **functional groups** are specific groups (moieties) of atoms or bonds within molecules that are responsible for the characteristic chemical reactions of those molecules. The same functional group will undergo the same or similar chemical reaction(s) regardless of the size of the molecule it is a part of. However, its relative reactivity can be modified by other functional groups nearby. The atoms of functional groups are linked to each other and to the rest of the molecule by covalent bonds. Any subgroup of atoms of a compound also may be called a **radical**, and if a covalent bond is broken homolytically, the resulting fragment radicals are referred as free radicals. Functional groups can also be charged, e.g. in carboxylate salts (-COO^-), which turns the molecule into a polyatomic ion or a complex ion.

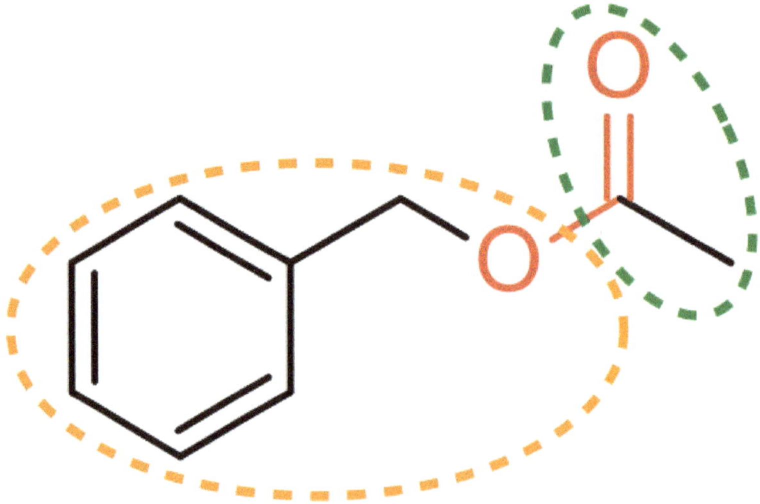

Benzyl acetate has an ester functional group (in red), an acetyl moiety (circled with dark green) and a benzyloxy moiety (circled with light orange). Other divisions can be made.

Complexation and solvation is also caused by specific interactions of functional groups. In the common rule of thumb "like dissolves like", it is the shared or mutually well-interacting functional groups which give rise to solubility. For example, sugar dissolves in water because both share the hydroxyl functional group (-OH) and hydroxyls interact strongly with each other.

Combining the names of functional groups with the names of the parent alkanes generates what is termed a systematic nomenclature for naming organic compounds. In traditional nomenclature, the first carbon atom after the carbon that attaches to the functional group is called the alpha carbon; the second, beta carbon, the third, gamma carbon, etc. If there is another functional group at a carbon, it may be named with the Greek letter, e.g., the gamma-amine in gamma-aminobutyric acid is on the third carbon of the carbon chain attached to the carboxylic acid group. IUPAC conventions call for numeric labeling of the position, e.g. 4-aminobutanoic acid. In traditional names various qualifiers are used to label isomers, for example isopropanol (IUPAC name: propan-2-ol) is an isomer is n-propanol (propan-1-ol).

Table of Common Functional Groups

The following is a list of common functional groups. In the formulas, the symbols R and R' usually denote an attached hydrogen, or a hydrocarbon side chain of any length, but may sometimes refer to any group of atoms.

Hydrocarbons

Functional groups, called hydrocarbyl, that contain only carbon and hydrogen, but vary in the number and order of double bonds. Each one differs in type (and scope) of reactivity.

Chemical class	Group	Formula	Structural Formulae	Prefix	Suffix	Example
Alkane	Alkyl	$R(CH_2)_nH$		alkyl-	-ane	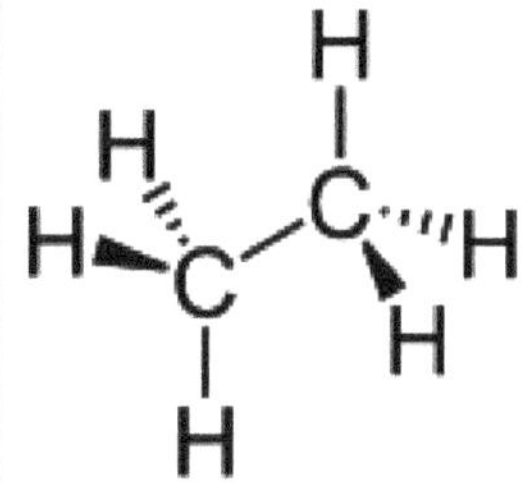 Ethane
Alkene	Alkenyl	$R_2C=CR_2$	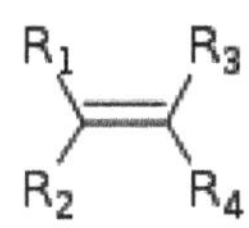	alkenyl-	-ene	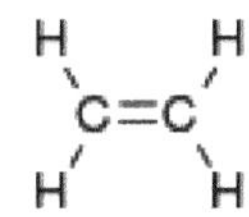 Ethylene *(Ethene)*
Alkyne	Alkynyl	RC≡CR’	R-C≡C-R’	alkynyl-	-yne	H-C≡C-H Acetylene *(Ethyne)*
Benzene derivative	Phenyl	RC_6H_5 RPh	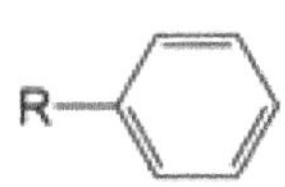	phenyl-	-benzene	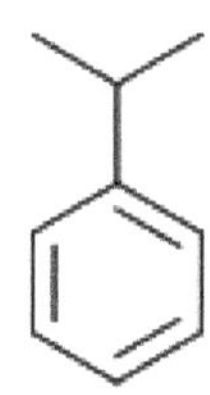 Cumene *(Isopropylbenzene)*

There are also a large number of branched or ring alkanes that have specific names, e.g., tert-butyl, bornyl, cyclohexyl, etc. Hydrocarbons may form charged structures: positively charged carbocations or negative carbanions. Carbocations are often named *-um*. Examples are tropylium and triphenylmethyl cations and the cyclopentadienyl anion.

Groups Containing Halogen

Haloalkanes are a class of molecule that is defined by a carbon–halogen bond. This bond can be relatively weak (in the case of an iodoalkane) or quite stable (as in the case of a fluoroalkane). In general, with the exception of fluorinated compounds, haloalkanes readily undergo nucleophilic substitution reactions or elimination reactions. The substitution on the carbon, the acidity of an adjacent proton, the solvent conditions, etc. all can influence the outcome of the reactivity.

Chemical class	Group	Formula	Structural Formula	Prefix	Suffix	Example
haloalkane	halo	RX	R-X	halo-	alkyl **halide**	 Chloroethane *(Ethyl chloride)*
fluoroalkane	fluoro	RF	R-F	fluoro-	alkyl **fluoride**	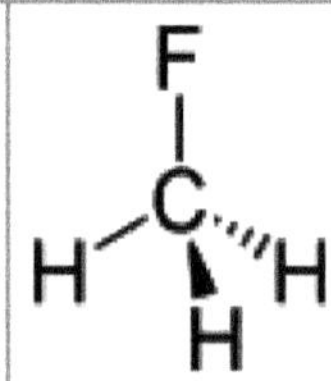 Fluoromethane *(Methyl fluoride)*
chloroalkane	chloro	RCl	R-Cl	chloro-	alkyl **chloride**	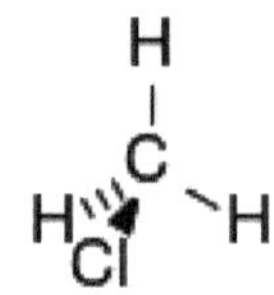 Chloromethane *(Methyl chloride)*
bromoalkane	bromo	RBr	R-Br	bromo-	alkyl **bromide**	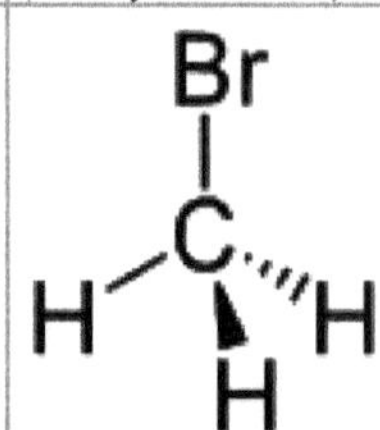 Bromomethane *(Methyl bromide)*
iodoalkane	iodo	RI	R-I	iodo-	alkyl **iodide**	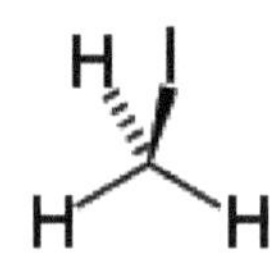 Iodomethane *(Methyl iodide)*

Groups Containing Oxygen

Compounds that contain C-O bonds each possess differing reactivity based upon the location and hybridization of the C-O bond, owing to the electron-withdrawing effect of sp-hybridized oxygen (carbonyl groups) and the donating effects of sp^2-hybridized oxygen (alcohol groups).

Chemical class	Group	Formula	Structural Formula	Prefix	Suffix	Example
Alcohol	Hydroxyl	ROH	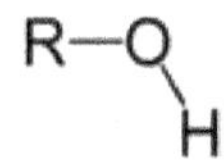	hydroxy-	**-ol**	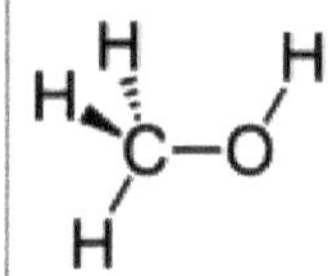 Methanol
Ketone	Carbonyl	RCOR'	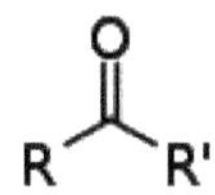	-oyl- (-COR') or oxo- (=O)	**-one**	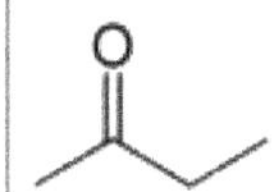 Butanone *(Methyl ethyl ketone)*
Aldehyde	Aldehyde	RCHO	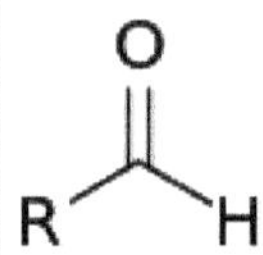	formyl- (-COH) or oxo- (=O)	**-al**	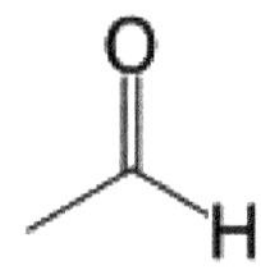 Acetaldehyde *(Ethanal)*
Acyl halide	Haloformyl	RCOX	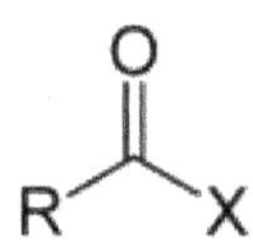	carbonofluoridoyl- carbonochloridoyl- carbonobromidoyl- carbonoiodidoyl-	-oyl **halide**	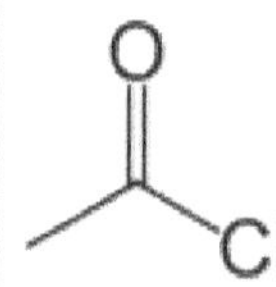 Acetyl chloride *(Ethanoyl chloride)*
Carbonate	Carbonate ester	ROCOOR	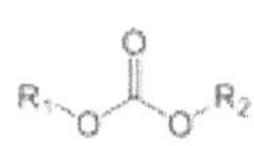	(alkoxycarbonyl) oxy-	alkyl **carbonate**	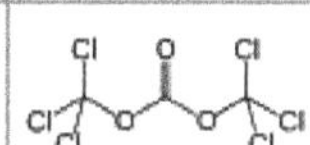 Triphosgene *(bis(trichloromethyl) carbonate)*
Carboxylate	Carboxylate	$RCOO^-$	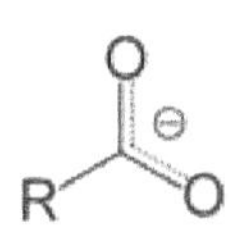 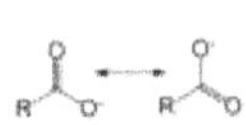	carboxy-	-oate	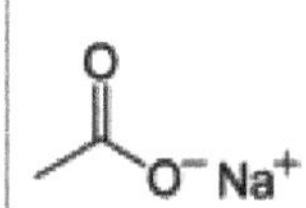 Sodium acetate *(Sodium ethanoate)*
Carboxylic acid	Carboxyl	RCOOH	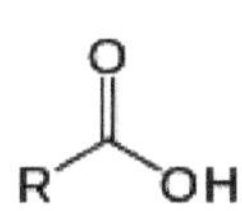	carboxy-	-oic **acid**	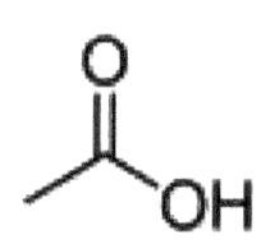 Acetic acid *(Ethanoic acid)*

Ester	Ester	RCOOR’	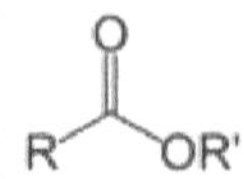	alkanoyloxy- or alkoxycarbonyl	alkyl alkan**oate**	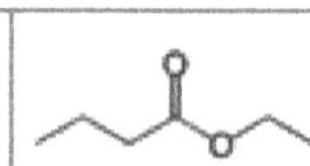 Ethyl butyrate *(Ethyl butanoate)*
Methoxy	Methoxy	$ROCH_3$	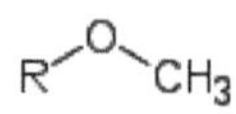	methoxy-		
Hydroperoxide	Hydroperoxy	ROOH	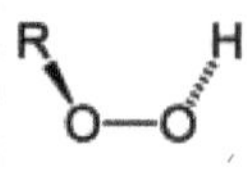	hydroperoxy-	alkyl **hydroperoxide**	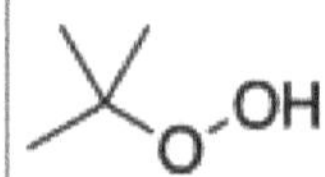 *tert*-Butyl hydroperoxide
Peroxide	Peroxy	ROOR	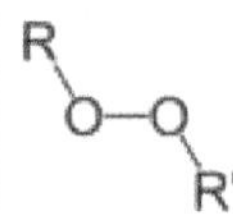	peroxy-	alkyl **peroxide**	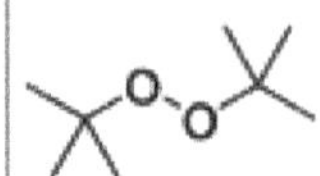 Di-tert-butyl peroxide
Ether	Ether	ROR’	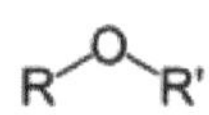	alkoxy-	alkyl **ether**	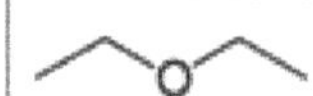 Diethyl ether *(Ethoxyethane)*
Hemiacetal	Hemiacetal	RCH(OR’)(OH)	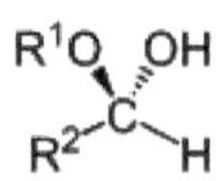	alkoxy -ol	-al alkyl **hemiacetal**	
Hemiketal	Hemiketal	RC(OR″)(OH)R’	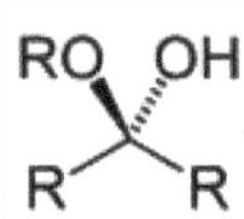	alkoxy -ol	-one alkyl **hemiketal**	
Acetal	Acetal	RCH(OR’)(OR”)	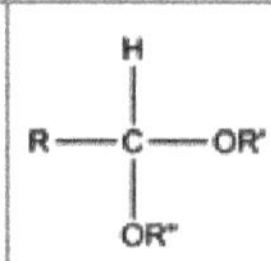	dialkoxy-	-al dialkyl **acetal**	
Ketal (or Acetal)	Ketal (or Acetal)	RC(OR’”)(OR’”)R’	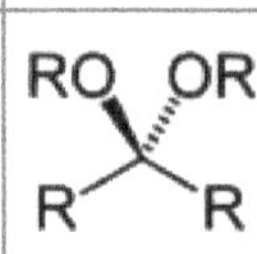	dialkoxy-	-one dialkyl **ketal**	
Orthoester	Orthoester	RC(OR’)(OR”)(OR’”)	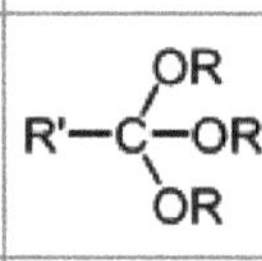	trialkoxy-		
Heterocycle	Methylenedioxy	PhOCOPh	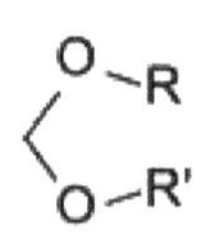	methylenedioxy-	-dioxole	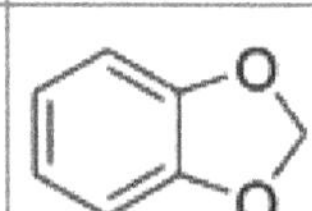 1,2-Methylenedioxybenzene *(1,3-Benzodioxole)*
Orthocarbonate ester	Orthocarbonate ester	C(OR)(OR’)(OR’”)(OR″)		tetralkoxy-	*tetraalkyl* **orthocarbonate**	

Groups Containing Nitrogen

Compounds that contain nitrogen in this category may contain C-O bonds, such as in the case of amides.

Chemical class	Group	Formula	Structural Formula	Prefix	Suffix	Example
Amide	Carboxamide	$RCONR_2$	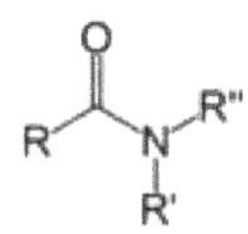	carboxamido- or carbamoyl-	-amide	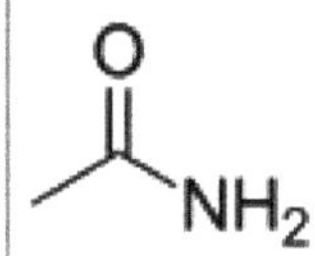 Acetamide *(Ethanamide)*
Amines	Primary amine	RNH_2	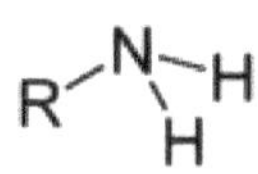	amino-	-amine	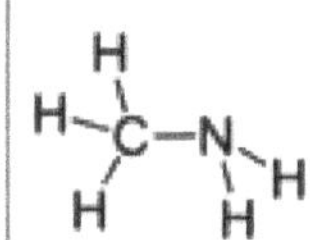 Methylamine *(Methanamine)*
	Secondary amine	R_2NH	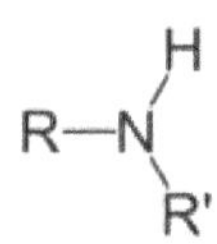	amino-	-amine	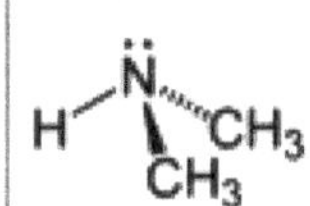 Dimethylamine
	Tertiary amine	R_3N	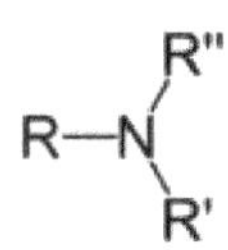	amino-	-amine	 Trimethylamine
	4° ammonium ion	R_4N^+	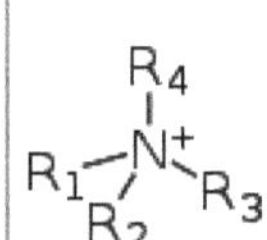	ammonio-	-ammonium	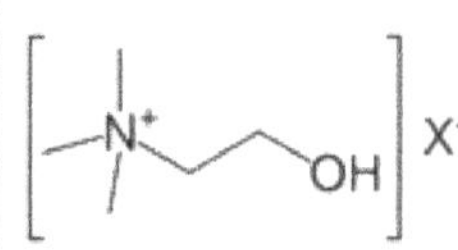 Choline

<table>
<tr><td rowspan="4">Imine</td><td>Primary ketimine</td><td>RC(=NH)R’</td><td>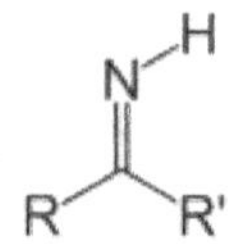</td><td>imino-</td><td>-imine</td><td></td></tr>
<tr><td>Secondary ketimine</td><td>RC(=NR”) R’</td><td>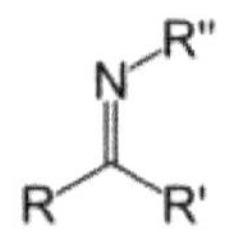</td><td>imino-</td><td>-imine</td><td></td></tr>
<tr><td>Primary aldimine</td><td>RC(=NH)H</td><td>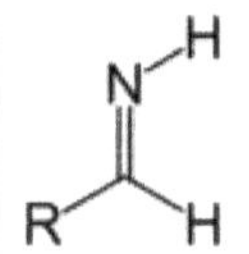</td><td>imino-</td><td>-imine</td><td>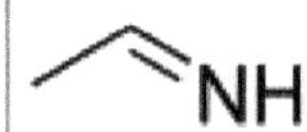
Ethanimine</td></tr>
<tr><td>Secondary aldimine</td><td>RC(=NR’)H</td><td>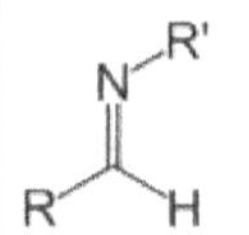</td><td>imino-</td><td>-imine</td><td></td></tr>
<tr><td>Imide</td><td>Imide</td><td>$(RCO)_2NR'$</td><td>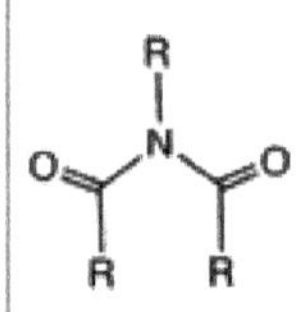</td><td>imido-</td><td>-imide</td><td>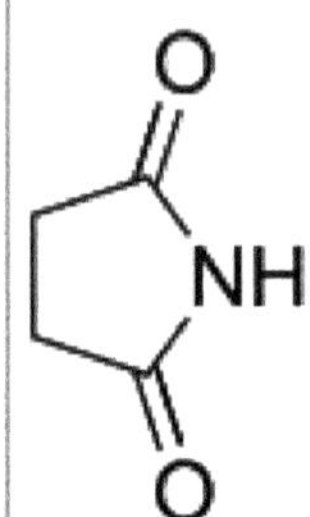
Succinimide
(Pyrrolidine-2,5-dione)</td></tr>
<tr><td>Azide</td><td>Azide</td><td>RN_3</td><td>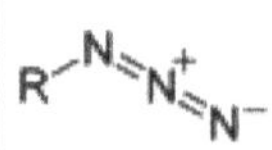</td><td>azido-</td><td>alkyl azide</td><td>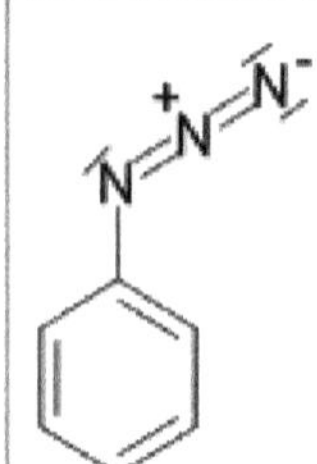
Phenyl azide
(Azidobenzene)</td></tr>
<tr><td>Azo compound</td><td>Azo (Diimide)</td><td>RN_2R'</td><td>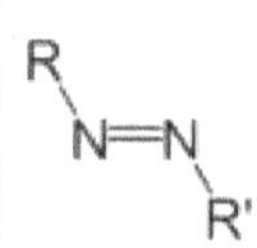</td><td>azo-</td><td>-diazene</td><td>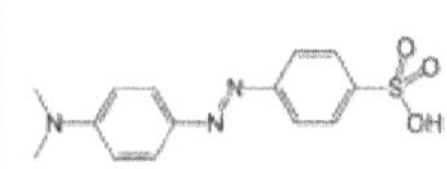
Methyl orange
(p-dimethylamino-azobenzenesulfonic acid)</td></tr>
</table>

Cyanates	Cyanate	ROCN	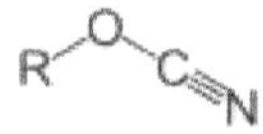	cyanato-	alkyl **cyanate**	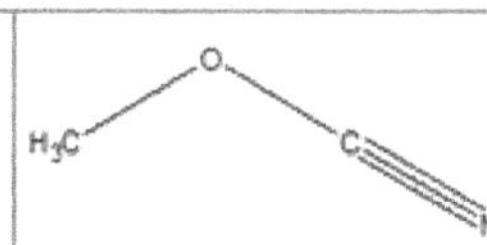 Methyl cyanate
	Isocyanate	RNCO	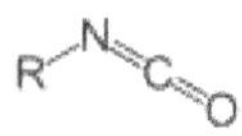	isocyanato-	alkyl **isocyanate**	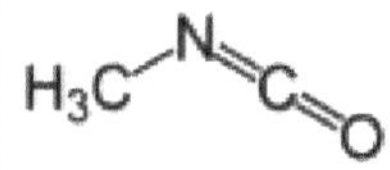 Methyl isocyanate
Nitrate	Nitrate	$RONO_2$	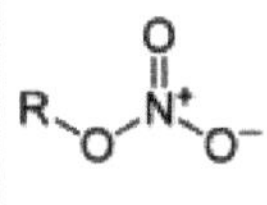	nitrooxy-, nitroxy-	alkyl **nitrate**	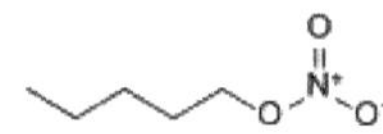 Amyl nitrate *(1-nitrooxypentane)*
Nitrile	Nitrile	RCN	R—≡N	cyano-	alkane**nitrile** alkyl **cyanide**	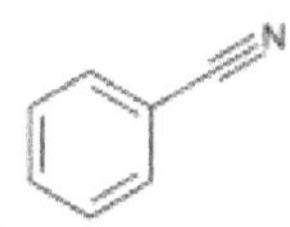 Benzonitrile *(Phenyl cyanide)*
	Isonitrile	RNC	$R{-}\overset{+}{N}{\equiv}C^-$	isocyano-	alkane**isonitrile** alkyl **isocyanide**	$H_3C{-}\overset{+}{N}{\equiv}C^-$ Methyl isocyanide
Nitrite	Nitrosooxy	RONO	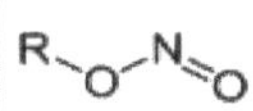	nitrosooxy-	alkyl **nitrite**	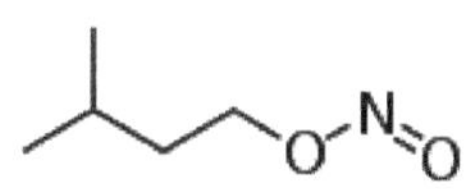 Isoamyl nitrite *(3-methyl-1-nitrosooxybutane)*
Nitro compound	Nitro	RNO_2	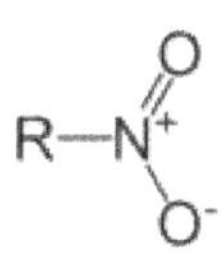	nitro-		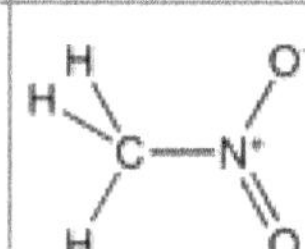 Nitromethane
Nitroso compound	Nitroso	RNO	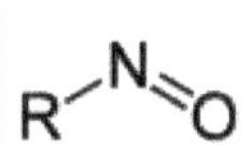	nitroso- (Nitrosyl-)		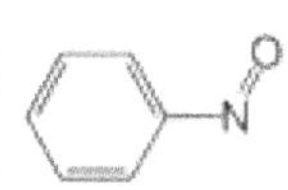 Nitrosobenzene
Oxime	Oxime	RCH=NOH	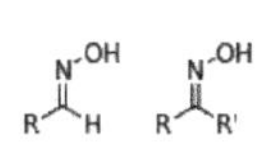 aldoxime ketoxime		Oxime	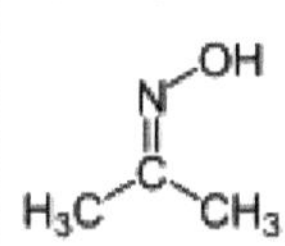 Acetone oxime *(2-Propanone oxime)*

Pyridine derivative	Pyridyl	RC_5H_4N	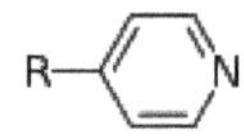 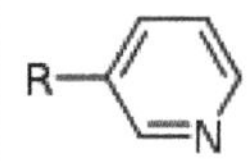 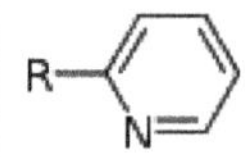	4-pyridyl (pyridin-4-yl) 3-pyridyl (pyridin-3-yl) 2-pyridyl (pyridin-2-yl)	-pyridine	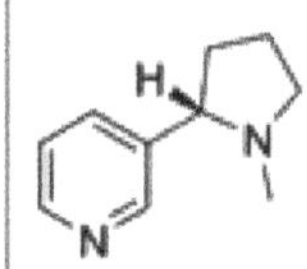 Nicotine

Groups Containing Sulfur

Compounds that contain sulfur exhibit unique chemistry due to their ability to form more bonds than oxygen, their lighter analogue on the periodic table. Substitutive nomenclature (marked as prefix in table) is preferred over functional class nomenclature (marked as suffix in table) for sulfides, disulfides, sulfoxides and sulfones.

Chemical class	Group	Formula	Structural Formula	Prefix	Suffix	Example
Thiol	Sulfhydryl	RSH	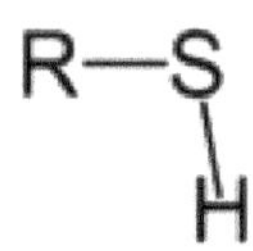	sulfanyl- (-SH)	**-thiol**	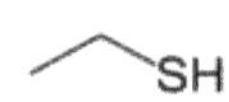 Ethanethiol
Sulfide (Thioether)	Sulfide	RSR’	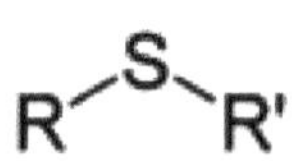	*substituent* sulfanyl- (-SR’)	di(*substituent*) **sulfide**	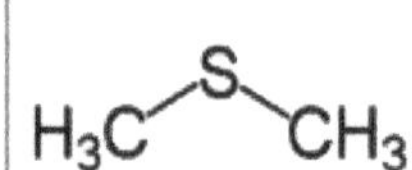 (Methylsulfanyl)methane (prefix) or Dimethyl sulfide (suffix)
Disulfide	Disulfide	RSSR’	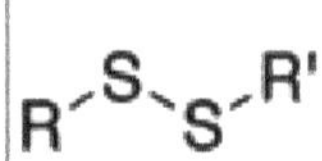	*substituent* disulfanyl- (-SSR’)	di(*substituent*) **disulfide**	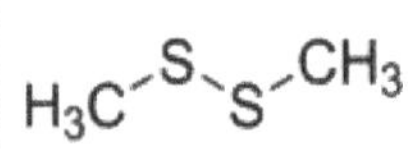 (Methyldisulfanyl) methane (prefix) or Dimethyl disulfide (suffix)
Sulfoxide	Sulfinyl	RSOR’	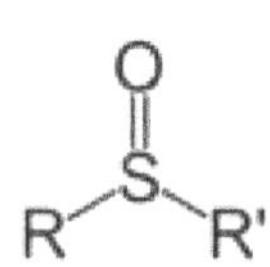	-sulfinyl- (-SOR’)	di(*substituent*) **sulfoxide**	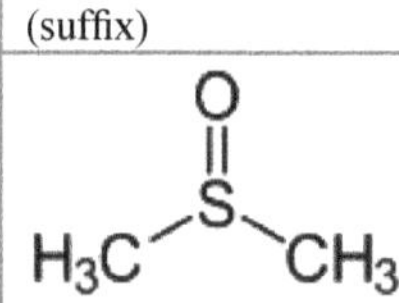 (Methanesulfinyl) methane (prefix) or Dimethyl sulfoxide (suffix)

Sulfone	Sulfonyl	RSO_2R'	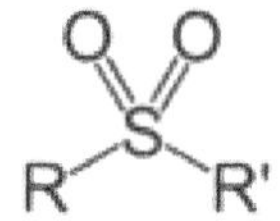	-sulfonyl- ($-SO_2R'$)	di(*substituent*) **sulfone**	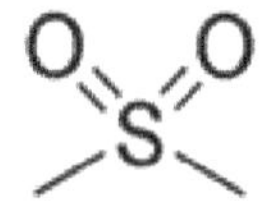 (Methanesulfonyl) methane (prefix) or Dimethyl sulfone (suffix)
Sulfinic acid	Sulfino	RSO_2H	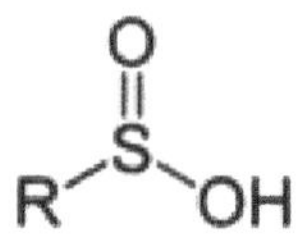	sulfino- ($-SO_2H$)	**-sulfinic acid**	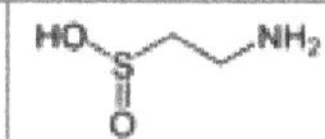 2-Aminoethanesulfinic acid
Sulfonic acid	Sulfo	RSO_3H	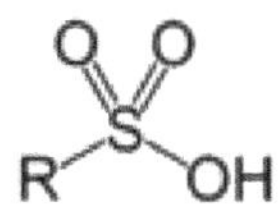	sulfo- ($-SO_3H$)	**-sulfonic acid**	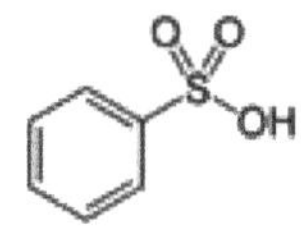 Benzenesulfonic acid
Thiocyanate	Thiocyanate	RSCN	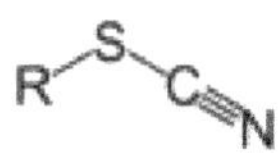	thiocyanato- (-SCN)	*substituent* **thiocyanate**	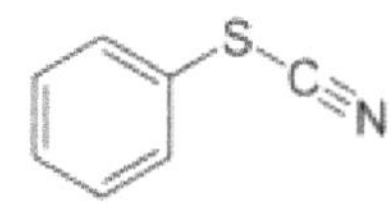 Phenyl thiocyanate
	Isothiocyanate	RNCS	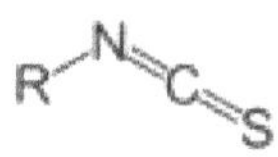	isothiocyanato- (-NCS)	*substituent* **isothiocyanate**	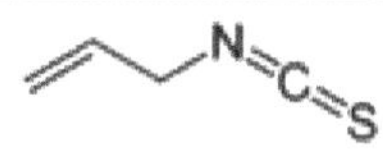 Allyl isothiocyanate
Thione	Carbonothioyl	RCSR'	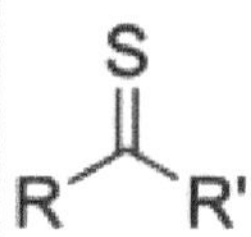	-thioyl- (-CSR') or sulfanylidene- (=S)	**-thione**	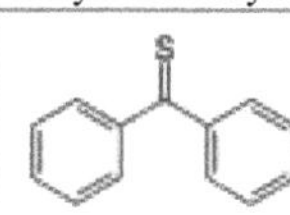 Diphenylmethanethione (Thiobenzophenone)
Thial	Carbonothioyl	RCSH	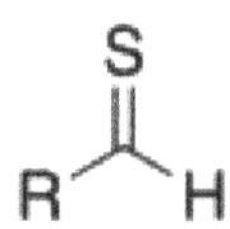	methanethioyl- (-CSH) or sulfanylidene- (=S)	**-thial**	
Thioester						
carbodithioic acid						

Groups Containing Phosphorus

Compounds that contain phosphorus exhibit unique chemistry due to their ability to form more bonds than nitrogen, their lighter analogues on the periodic table.

Chemical class	Group	Formula	Structural Formula	Prefix	Suffix	Example
Phosphine (Phosphane)	Phosphino	R_3P	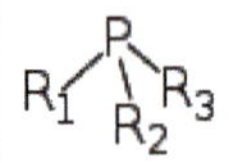	phosphanyl-	-phosphane	 Methylpropyl phosphane
Phosphonic acid	Phosphono	RP(=O)(OH)2	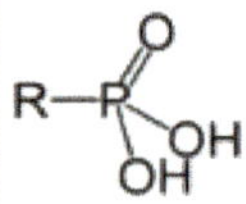	phosphono-	*substituent* **phosphonic acid**	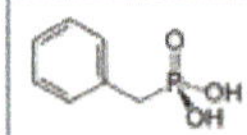 Benzylphosphonic acid
Phosphate	Phosphate	ROP(=O)(OH)2	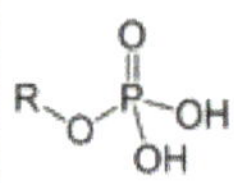	phosphonooxy- or *O*-phosphono- (phospho-)	*substituent* **phosphate**	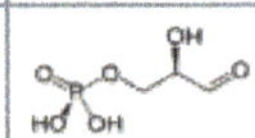 Glyceraldehyde 3-phosphate (suffix) 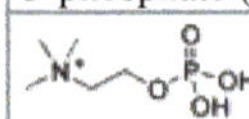 *O*-Phosphonocholine (prefix) (Phosphocholine)
Phosphodiester	Phosphate	$HOPO(OR)_2$	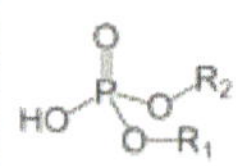	[(alkoxy) hydroxyphosphoryl] oxy- or *O*-[(alkoxy) hydroxyphosphoryl]-	di(*substituent*) hydrogen **phosphate** or phosphoric acid di(*substituent*) **ester**	DNA *O* [(2Guanidinoethoxy) hydroxyphosphoryl] ɪserine (prefix) (Lombricine)

Groups Containing Boron

Compounds containing boron exhibit unique chemistry due to their having partially filled octets and therefore acting as Lewis acids.

Chemical class	Group	Formula	Structural Formula	Prefix	Suffix	Example
Boronic acid	Borono	$RB(OH)_2$	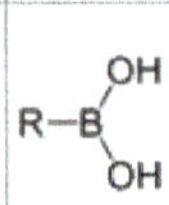	Borono-	*substituent* **boronic acid**	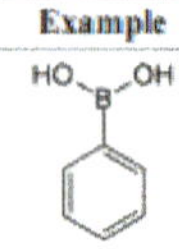 Phenylboronic acid
Boronic ester	Boronate	$RB(OR)_2$	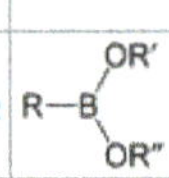	O-[bis(alkoxy)alkylboronyl]-	*substituent* **boronic acid** di(*substituent*) **ester**	
Borinic acid	Borino	R_2BOH	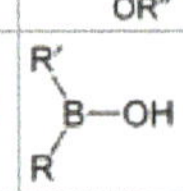	Hydroxyborino-	di(*substituent*) **borinic acid**	
Borinic ester	Borinate	R_2BOR	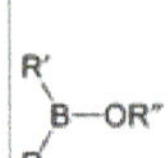	O-[alkoxydialkylboronyl]-	di(*substituent*) **borinic acid** *substituent* **ester**	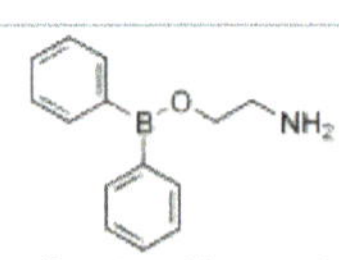 Diphenylborinic acid 2-aminoethyl ester (2-Aminoethoxydiphenyl borate)

Names of Radicals or Moieties

These names are used to refer to the moieties themselves or to radical species, and also to form the names of halides and substituents in larger molecules.

When the parent hydrocarbon is unsaturated, the suffix ("-yl", "-ylidene", or "-ylidyne") replaces "-ane" (e.g. "ethane" becomes "ethyl"); otherwise, the suffix replaces only the final "-e" (e.g. "ethyne" becomes "ethynyl").

Note that when used to refer to moieties, multiple single bonds differ from a single multiple bond. For example, a methylene bridge (methanediyl) has two single bonds, whereas a methylene group (methylidene) has one double bond. Suffixes can be combined, as in methylidyne (triple bond) vs. methylylidene (single bond and double bond) vs. methanetriyl (three single bonds).

There are some retained names, such as methylene for methanediyl, 1,x-phenylene for phenyl-1,x-diyl (where x is 2, 3, or 4),carbyne for methylidyne, and trityl for triphenylmethyl.

Chemical class	Group	Formula	Structural Formula	Prefix	Suffix	Example
Single bond		R•		Ylo-[6]	-yl	Methyl group Methyl radical
Double bond		R:		?	-ylidene	Methylidene
Triple bond		R⁝		?	-ylidyne	Methylidyne
Carboxylic acyl radical	Acyl	R−C(=O)•		?	-oyl	Acetyl

REFERENCES

1. Compendium of Chemical Terminology (IUPAC "Gold Book") http://goldbook.iupac.org/F02555.html
2. March, Jerry (1985), Advanced Organic Chemistry: Reactions, Mechanisms, and Structure (3rd ed.), New York: Wiley, ISBN 0-471-85472-7
3. Brown, Theodore (2002). Chemistry : the central science. Upper Saddle River, NJ: Prentice Hall. p. 1001. ISBN 0130669970.
4. Moss, G. P.; W.H. Powell. RC-81.1.1. Monovalent radical centers in saturated acyclic and monocyclic hydrocarbons, and the mononuclear EH4 parent hydrides of the carbon family". IUPAC Recommendations 1993. Department of Chemistry, Queen Mary University of London. Archived from the original on 9 February 2015. Retrieved 25 February 2015.
5. R-2. 5 Substituent Prefix Names Derived from Parent Hydrides". IUPAC. 1993. section P-56.2.1
6. Revised Nomenclature for Radicals, Ions, Radical Ions and Related Species (IUPAC Recommendations 1993: RC-81.3. Multiple radical centers)"

CHAPTER 3

Nomenclature of Organic Chemistry

Chapter Contents

- **Introduction**
- **Basic Principles**
- **Alkanes**
- **Alkenes and Alkynes**
- **Functional Groups**
- **Common Nomenclature – Trivial Names**
- **Ions**
- **Polycyclic Compound**
- **Bicyclic Molecule**
- **Spiro Compound**

The increasingly large number of organic compounds identified with each passing day, together with the fact that many of these compounds are isomers of other compounds, requires that a systematic nomenclature system be developed. Just as each distinct compound has a unique molecular structure which can be designated by a structural formula, each compound must be given a characteristic and unique name. As organic chemistry grew and developed, many compounds were given trivial names, which are now commonly used and recognized.

INTRODUCTION

In chemical nomenclature, the **IUPAC nomenclature of organic chemistry** is a systematic method of naming organic chemical compounds as recommended by the International Union of Pure and Applied Chemistry (IUPAC). It is published in the *Nomenclature of Organic Chemistry* (informally called the Blue Book). Ideally, every possible organic compound should have a name from which an unambiguous structural formula can be created. There is also an IUPAC nomenclature of inorganic chemistry.

For ordinary communication, to spare a tedious description, the official IUPAC naming recommendations are not always followed in practice, except when it is necessary to give an unambiguous and absolute definition to a compound, or when the IUPAC name is simpler (e.g. ethanol instead of ethyl alcohol). Otherwise the common or trivial name may be used, often derived from the source of the compound (see below). In addition, very long names may be less concise than structural formulae.

BASIC PRINCIPLES

In chemistry, a number of prefixes, suffixes and infixes are used to describe the type and position of functional groups in the compound.

The steps for naming an organic compound are:

- Identification of the parent hydrocarbon chain. This chain must obey the following rules, in order of precedence:
- It should have the maximum number of substituents of the suffix functional group. By suffix, it is meant that the parent functional group should have a suffix, unlike halogen substituents. If more than one functional group is present, the one with highest precedence should be used.
- It should have the maximum number of multiple bonds
- It should have the maximum number of single bonds.
- It should have the maximum length.
- Identification of the parent functional group, if any, with the highest order of precedence.
- Identification of the side-chains. *Side chains are the carbon chains that are not in the parent chain, but are branched off from it.*
- Identification of the remaining functional groups, if any, and naming them by their ionic prefixes (such as hydroxy for -OH, oxy for =O, oxyalkane for O-R, etc.).Different side-chains and functional groups will be grouped together in **alphabetical** order. (The prefixes di-, tri-, etc. are not taken into consideration for grouping alphabetically. For example, ethyl comes before dihydroxy or dimethyl, as the e in ethyl precedes the h in dihydroxy and the m in dimethyl alphabetically. The di is not considered in either case). When both side chains and secondary functional groups are present, they should be written mixed together in one group rather than in two separate groups.
- Identification of double/triple bonds.
- Numbering of the chain. This is done by first numbering the chain in both directions (left to right and right to left), and then choosing the numbering which follows these rules, in order of precedence
- Has the lowest-numbered locant (or locants) for the suffix functional group. Locants are the numbers on the carbons to which the substituent is directly attached.
- Has the lowest-numbered locants for multiple bonds (The locant of a multiple bond is the number of the adjacent carbon with a lower number).
- Has the lowest-numbered locants for prefixes.
- Numbering of the various substituents and bonds with their locants. If there is more than one of the same type of substituent/double bond, a prefix is added

showing how many there are (di – 2 tri – 3 tetra – 4 then as for the number of carbons below with ‹a› added)

The numbers for that type of side chain will be grouped in ascending order and written before the name of the side-chain. If there are two side-chains with the same alpha carbon, the number will be written twice. Example: 2,2,3-trimethyl- . If there are both double bonds and triple bonds, en (double bond) is written before yne (triple bond). When the main functional group is a terminal functional group (a group which can exist only at the end of a chain, like formyl and carboxyl groups), there is no need to number it.

- Arrangement in this form: Group of side chains and secondary functional groups with numbers made in step 3 + prefix of parent hydrocarbon chain (eth, meth) + double/triple bonds with numbers (or ane) + primary functional group suffix with numbers.Wherever it says with numbers , it is understood that between the word and the numbers, the prefix(di-, tri-) is used.
- Adding of punctuation:
- Commas are put between numbers (2 5 5 becomes 2,5,5)
- Hyphens are put between a number and a letter (2 5 5 trimethylheptane becomes 2,5,5-trimethylheptane)
- Successive words are merged into one word (trimethyl heptane becomes trimethylheptane)Note: IUPAC uses one-word names throughout. This is why all parts are connected.

The finalized name should look like this: #,#-di<*side chain*>-#-<*secondary functional group*>-#-<*side chain*>-#,#,#-tri<*secondary functional group*><*parent chain prefix*><*If all bonds are single bonds, use "ane"*>-#,#-di<*double bonds*>-#-<*triple bonds*>-#-<*primary functional group*>Note: # is used for a number. The group secondary functional groups and side chains may not look the same as shown here, as the side chains and secondary functional groups are arranged alphabetically. The di- and tri- have been used just to show their usage. (di- after #,#, tri- after #,#,#, etc.)

Example

Here is a sample molecule with the parent carbons numbered:

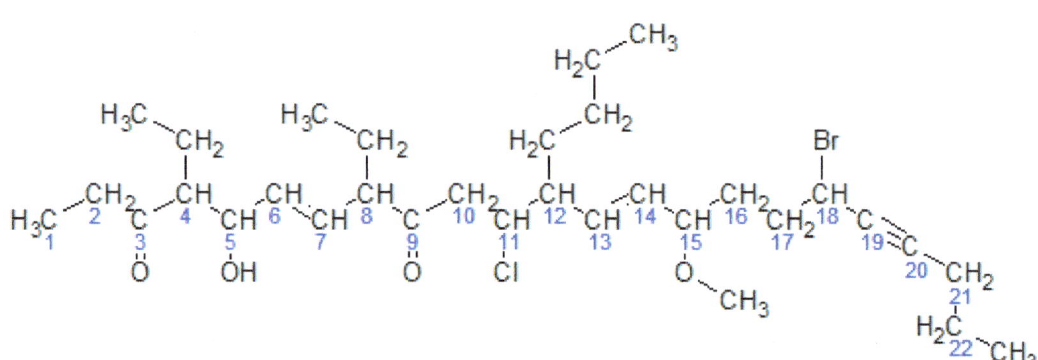

For simplicity, here is an image of the same molecule, where the hydrogens in the parent chain are removed and the carbons are shown by their numbers:

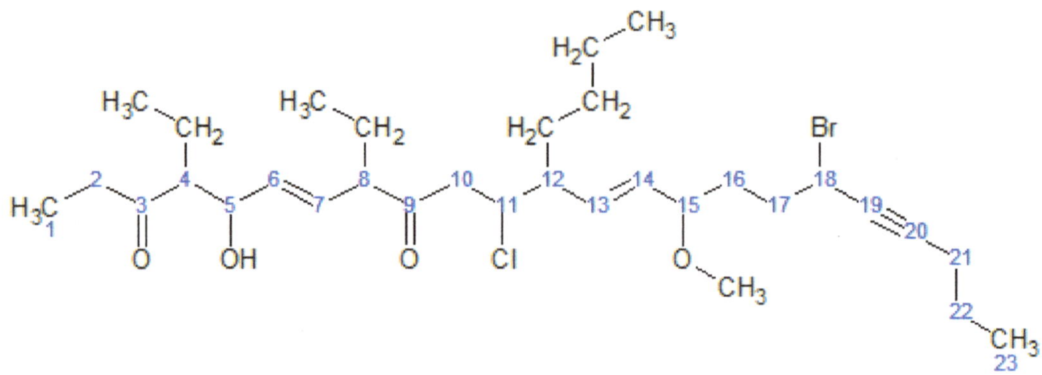

Now, following the above steps:

- The parent hydrocarbon chain has 23 carbons. It is called **tricosa-**.
- The functional groups with the highest precedence are the two ketone groups.
- The groups are on carbon atoms 3 and 9. As there are two, we write **3,9-dione**.
- The numbering of the molecule is based on the ketone groups. When numbering from left to right, the ketone groups are numbered 3 and 9. When numbering from right to left, the ketone groups are numbered 15 and 21. 3 is less than 15, therefore the ketones are numbered 3 and 9. The **smaller number** is always used, **not the sum** of the constituents numbers.
- The side chains are: an ethyl- at carbon 4, an ethyl- at carbon 8, and a butyl- at carbon 12.Note:The -O-CH_3 at carbon atom 15 is not a side chain, but it is a methoxy functional group
- There are two ethyl- groups. They are combined to create, **4,8-diethyl**.
- The side chains are grouped like this: **12-butyl-4,8-diethyl**. (But this is not the final grouping, as functional groups may be added in between.)
- The secondary functional groups are: a hydroxy- at carbon 5, a chloro- at carbon 11, a methoxy- at carbon 15, and a bromo- at carbon 18. Grouped with the side chains, this gives **18-bromo-12-butyl-11-chloro-4,8-diethyl-5-hydroxy-15-methoxy**
- There are two double bonds: one between carbons 6 and 7, and one between carbons 13 and 14. They would be called "6,13-diene", but the presence of alkynes switches it to **6,13-dien**. There is one triple bond between carbon atoms 19 and 20. It will be called **19-yne**
- The arrangement (with punctuation) is: **18-bromo-12-butyl-11-chloro-4,8-diethyl-5-hydroxy-15-methoxytricosa-6,13-dien-19-yne-3,9-dione**
- Finally, due to Cis-trans isomerism, we have to specify the relative orientation of functional groups around each double bond. For this example, we have **(6*E*,13*E*)**

The final name is **(6*E*,13*E*)-18-bromo-12-butyl-11-chloro-4,8-diethyl-5-hydroxy-15-methoxytricosa-6,13-dien-19-yne-3,9-dione**.

ALKANES

Straight-chain alkanes take the suffix "-ane and are prefixed depending on the number of carbon atoms in the chain, following standard rules. The first few are:

Number of carbons	1	2	3	4	5	6	7	8	9	10	11	12	13	14	15	20	30	40	50
Prefix	Meth	Eth	Prop	But	Pent	Hex	Hept	Oct	Non	Dec	Undec	Dodec	Tridec	Tetradec	Pentadec	Eicos	Triacont	Tetracont	Pentacont

For example, the simplest alkane is CH_4 methane, and the nine-carbon alkane $CH_3(CH_2)_7CH_3$ is named nonane. The names of the first four alkanes were derived from methanol, ether, propionic acid and butyric acid, respectively. The rest are named with a Greek numeric prefix, with the exceptions of nonane which has a Latin prefix, and undecane and tridecane which have mixed-language prefixes.

Cyclic alkanes are simply prefixed with cyclo- : for example, C_4H_8 is cyclobutane (not to be confused with butene) and $C_6H_{12}$is cyclohexane (not to be confused with hexene).

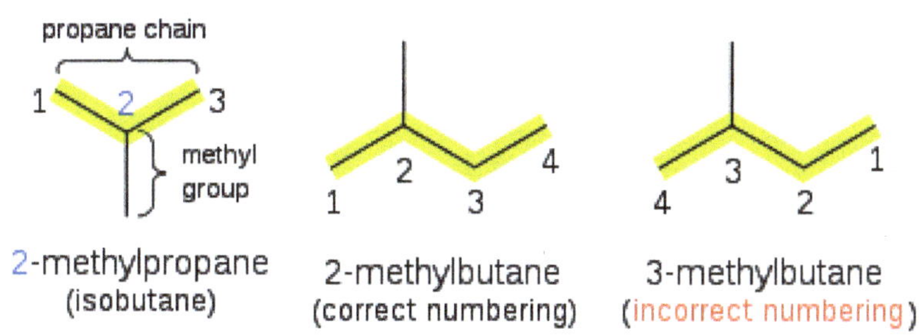

Branched alkanes are named as a straight-chain alkane with attached alkyl groups. They are prefixed with a number indicating the carbon the group is attached to, counting from the end of the alkane chain. For example, $(CH_3)_2CHCH_3$, commonly known as isobutane, is treated as a propane chain with a methyl group bonded to the middle (2) carbon, and given the systematic name 2-methylpropane. However, although the name 2-methylpropane *could* be used, it is easier and more logical to call it simply methylpropane – the methyl group could not possibly occur on any of the other carbon atoms (that would lengthen the chain and result in butane, not propane) and therefore the use of the number 2 is unnecessary.

If there is ambiguity in the position of the substituent, depending on which end of the alkane chain is counted as 1 , then numbering is chosen so that the smaller number is used. For example, $(CH_3)_2CHCH_2CH_3$ (isopentane) is named 2-methylbutane, not 3-methylbutane.

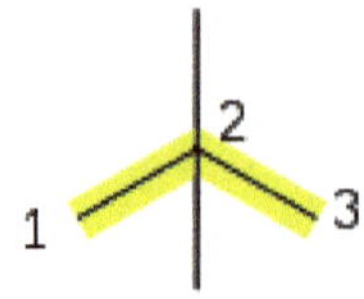

2,2-dimethylpropane
(neopentane)

If there are multiple side-branches of the same size alkyl group, their positions are separated by commas and the group prefixed with di-, tri-, tetra-, etc., depending on the number of branches. For example, $C(CH_3)_4$ (neopentane) is named 2,2-dimethylpropane. If there are different groups, they are added in alphabetical order, separated by commas or hyphens: . The longest possible main alkane chain is used; therefore 3-ethyl-4-methylhexane instead of 2,3-diethylpentane, even though these describe equivalent structures. The di-, tri- etc. prefixes are ignored for the purpose of alphabetical ordering of side chains (e.g. 3-ethyl-2,4-dimethylpentane, not 2,4-dimethyl-3-ethylpentane).

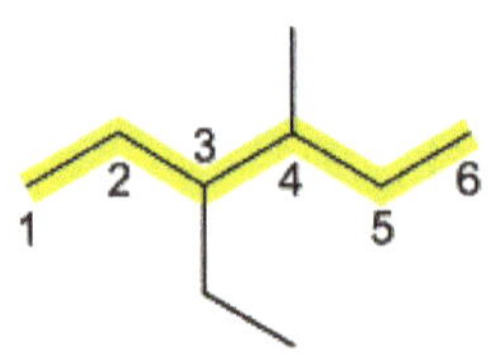

3-ethyl-4-methylhexane

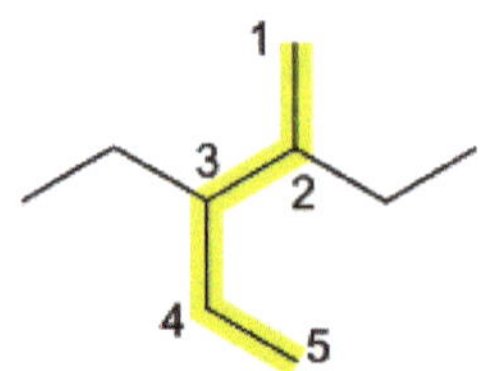

2,3-diethylpentane
(incorrect)

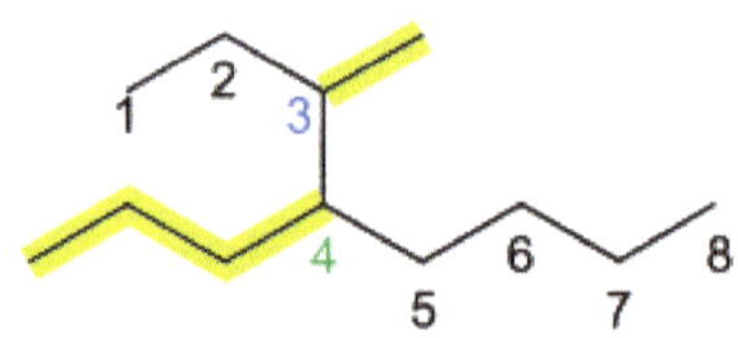

3-methyl-4-propyloctane

ALKENES AND ALKYNES

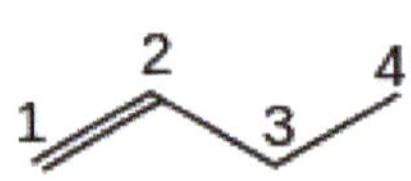

but-1-ene

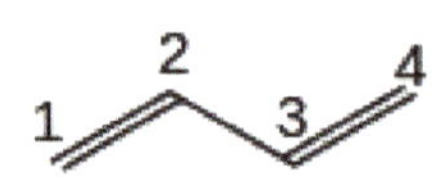

buta-1,3-diene

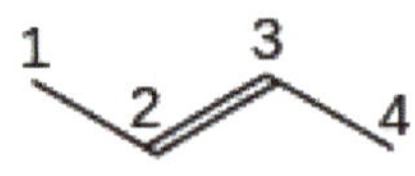

trans-but-2-ene
(E)-but-2-ene

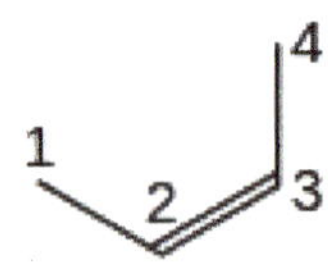

cis-but-2-ene
(Z)-but-2-ene

Alkenes are named for their parent alkane chain with the suffix "-ene" and an infixed number indicating the position of the carbon with the lower number for each double bond in the chain: $CH_2{=}CHCH_2CH_3$ is but-1-ene. Multiple double bonds take the form -diene, -triene, etc., with the size prefix of the chain taking an extra a : $CH_2{=}CHCH{=}CH_2$ is buta-1,3-diene. Simple cis and trans isomers may be indicated with a prefixed *cis-* or *trans-*: *cis*-but-2-ene, *trans*-but-2-ene. However, *cis-* and *trans-* are *relative* descriptors. It is IUPAC convention to describe all alkenes using *absolute* descriptors of *Z*- (same side) and *E*- (opposite) with the Cahn–Ingold–Prelog priority rules.

ethyne (acetylene) propyne (methylacetylene)

Alkynes are named using the same system, with the suffix "-yne" indicating a triple bond: ethyne (acetylene), propyne (methylacetylene).

FUNCTIONAL GROUPS

Alcohols

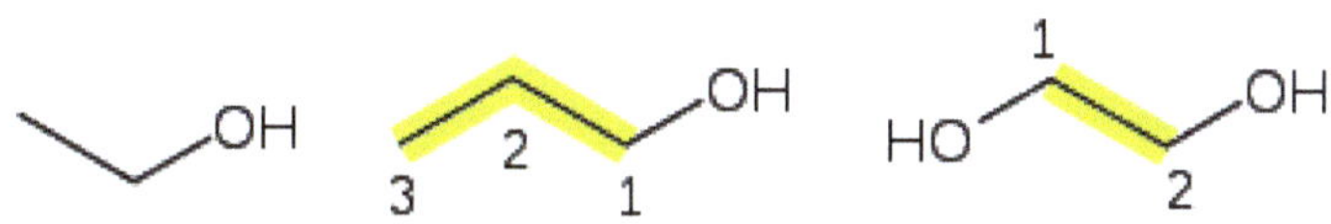

ethanol (ethyl alcohol) propan-1-ol (propyl alcohol)*n*- ethane-1,2-diol (ethylene glycol)

Alcohols (R-OH) take the suffix "-ol" with an infix numerical bonding position: $CH_3CH_2CH_2OH$ is propan-1-ol. The suffixes -diol, -triol, -tetraol, etc., are used for multiple -OH groups: Ethylene glycol CH_2OHCH_2OH is ethane-1,2-diol.

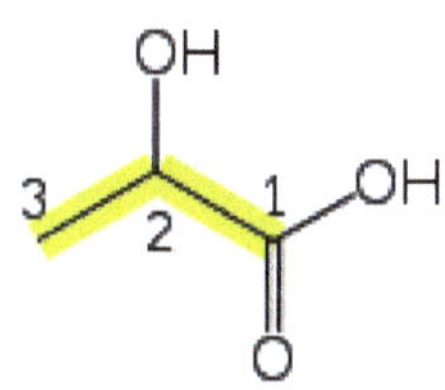

2-hydroxypropanoic acid

If higher precedence functional groups are present (see *order of precedence*, below), the prefix "hydroxy" is used with the bonding position: $CH_3CHOHCOOH$ is 2-hydroxypropanoic acid.

Halogens (alkyl halides)

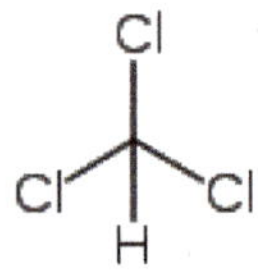

trichloromethane
(chloroform)

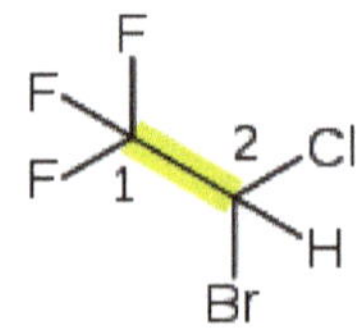

2-bromo-2-chloro-1,1,1,-trifluoroethane
(halothane)

Halogen functional groups are prefixed with the bonding position and take the form fluoro-, chloro-, bromo-, iodo-, etc., depending on the halogen. Multiple groups are dichloro-, trichloro-, etc., and dissimilar groups are ordered alphabetically as before. For example, $CHCl_3$ (chloroform) is trichloromethane. The anesthetic Halothane ($CF_3CHBrCl$) is 2-bromo-2-chloro-1,1,1-trifluoroethane.

Ketones

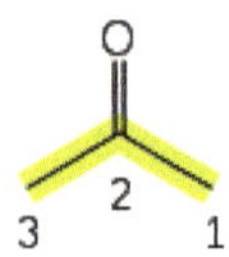

propan-2-one
(acetone)

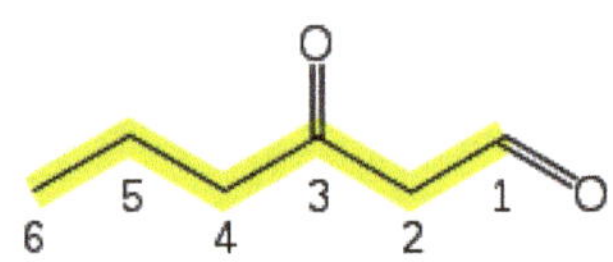

3-oxohexanal

In general ketones (R-CO-R) take the suffix "-one" (pronounced *own*, not *won*) with an infix position number: $CH_3CH_2CH_2COCH_3$ is pentan-2-one. **If a higher precedence suffix is in use, the prefix oxo- is used:** $CH_3CH_2CH_2COCH_2CHO$ is 3-oxohexanal.

Aldehydes

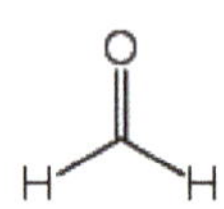

methanal
(formaldehyde)

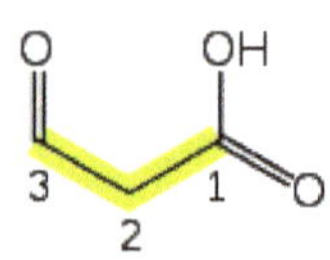

3-oxopropanoic acid

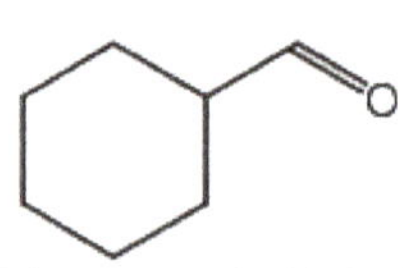

cyclohexanecarbaldehyde

Aldehydes (R-CHO) take the suffix "-al". If other functional groups are present, the chain is numbered such that the aldehyde carbon is in the "1" position, unless functional groups of higher precedence are present.

If a prefix form is required, "oxo-" is used (as for ketones), with the position number indicating the end of a chain: $CHOCH_2COOH$ is 3-oxopropanoic acid. If the carbon in the carbonyl group cannot be included in the attached chain (for instance in the case of cyclic aldehydes), the prefix "formyl-" or the suffix "**-carbaldehyde**" is used: $C_6H_{11}CHO$ is cyclohexanecarbaldehyde. If an aldehyde is attached to a benzene and is the main functional group, the suffix becomes benzaldehyde.

Carboxylic Acids

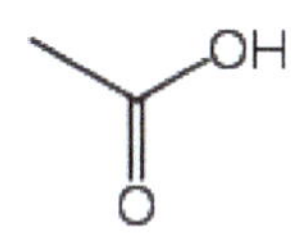

ethanoic acid
(acetic acid)

In general carboxylic acids are named with the suffix *-oic acid* (etymologically a back-formation from benzoic acid). Similar to aldehydes, they take the 1 position on the parent chain, but do not have their position number indicated. For example, $CH_3CH_2CH_2CH_2COOH$ (valeric acid) is named pentanoic acid. For common carboxylic acids some traditional names such as acetic acid are in such widespread use they are considered retained IUPAC names, although systematic names such as ethanoic acid are also acceptable. For carboxylic acids attached to a benzene ring such as Ph-COOH, these are named as benzoic acid or its derivatives.

If there are multiple carboxyl groups on the same parent chain, the suffix "-carboxylic acid" can be used (as -dicarboxylic acid, -tricarboxylic acid, etc.). In these cases, the carbon in the carboxyl group does *not* count as being part of the main alkane chain. The same is true for the prefix form, "carboxyl-". Citric acid is one example; it is named 2-hydroxypropane- 1,2,3-tricarboxylic acid, rather than 3-carboxy-3-hydroxypentanedioic acid.

Ethers

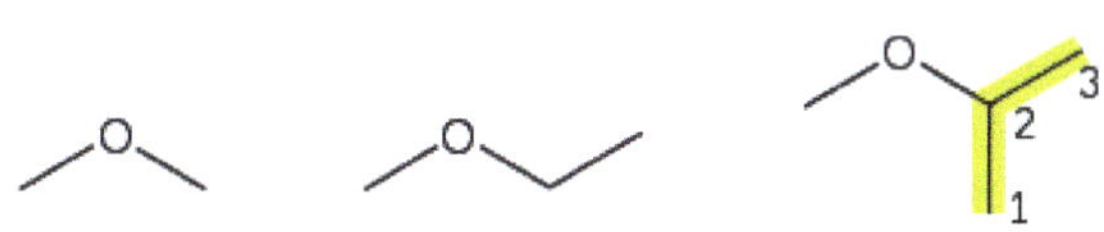

methoxymethane (dimethyl ether) | methoxyethane (ethyl methyl ether) | 2-methoxypropane (isopropyl methyl ether)

Ethers (R-O-R) consist of an oxygen atom between the two attached carbon chains. The shorter of the two chains becomes the first part of the name with the -ane suffix changed to -oxy, and the longer alkane chain becomes the suffix of the name of the ether. Thus, CH_3OCH_3 is methoxymethane, and $CH_3OCH_2CH_3$ is methoxyethane (*not* ethoxymethane). If the oxygen is not attached to the end of the main alkane chain, then the whole shorter alkyl-plus-ether group is treated as a side-chain and prefixed with its bonding position on the main chain. Thus $CH_3OCH(CH_3)_2$ is 2-methoxypropane.

Esters

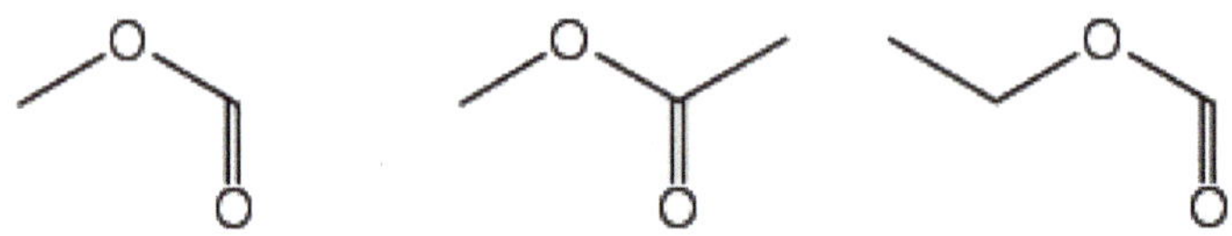

methyl methanoate (formate) | methyl ethanoate (acetate) | ethyl methanoate (ethyl formate)

Esters (R-CO-O-R') are named as alkyl derivatives of carboxylic acids. The alkyl (R') group is named first. The R-CO-O part is then named as a separate word based on the carboxylic acid name, with the ending changed from *-oic acid* to *-oate*. For example, $CH_3CH_2CH_2CH_2COOCH_3$ is *methyl pentanoate*, and $(CH_3)_2CHCH_2CH_2COOCH_2CH_3$ is *ethyl 4-methylpentanoate*. For esters such as ethyl acetate ($CH_3COOCH_2CH_3$), ethyl formate ($HCOOCH_2CH_3$) or dimethyl phthalate that are based on common acids, IUPAC recommends use of these established names, called retained names. The *-oate* changes to *-ate*. Some simple examples, named both ways, are shown in the figure above.

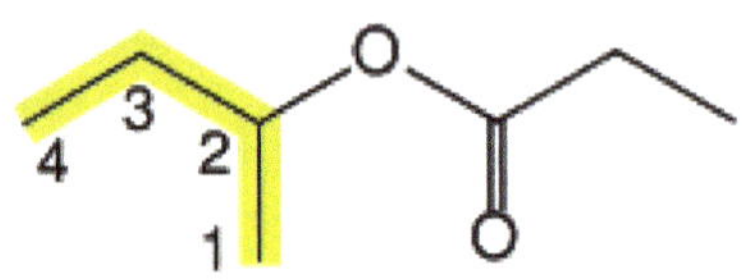

but-2-yl propanoate

If the alkyl group is not attached at the end of the chain, the bond position to the ester group is infixed before "-yl": $CH_3CH_2CH(CH_3)OOCCH_2CH_3$ may be called but-2-yl propanoate or but-2-yl propionate.

Amines and Amides

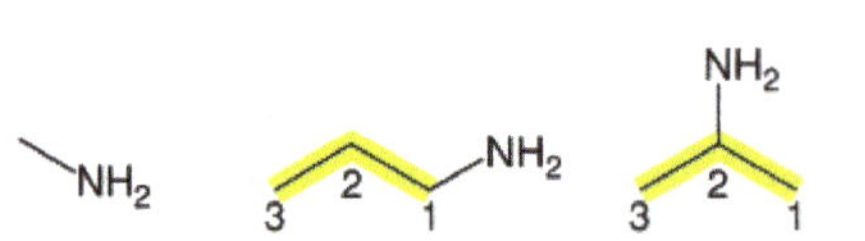

methanamine (methylamine) | propan-1-amine (*n*-propyl amine) | propan-2-amine (isopropyl amine)

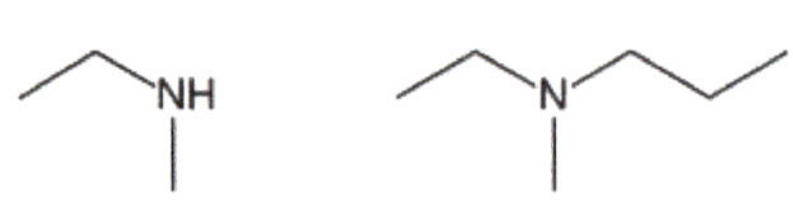

N-methylethanamine | *N*-ethyl-*N*-methylpropanamine

Amines ($R\text{-}NH_2$) are named for the attached alkane chain with the suffix "-amine" (e.g. CH_3NH_2 methanamine). If necessary, the bonding position is infixed: $CH_3CH_2CH_2NH_2$ propan-1-amine, $CH_3CHNH_2CH_3$ propan-2-amine. The prefix form is "amino-".

For secondary amines (of the form R-NH-R), the longest carbon chain attached to the nitrogen atom becomes the primary name of the amine; the other chain is prefixed as an alkyl group with location prefix given as an italic *N*: $CH_3NHCH_2CH_3$ is *N*-methylethanamine. Tertiary amines (R-NR-R) are treated similarly: $CH_3CH_2N(CH_3)CH_2CH_2CH_3$ is *N*-ethyl-*N*-methylpropanamine. Again, the substituent groups are ordered alphabetically.

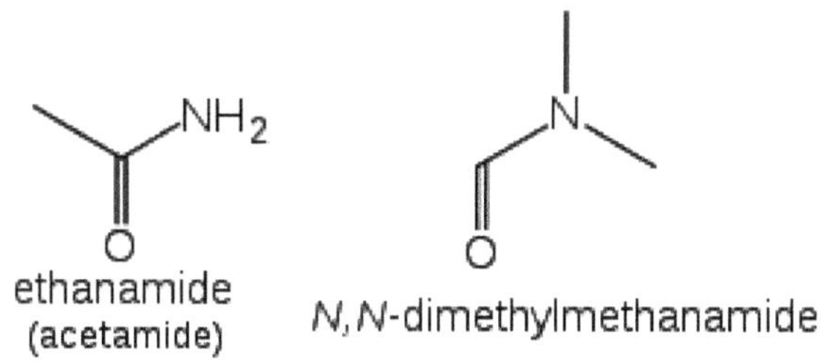

Amides ($R-CO-NH_2$) take the suffix "-amide", or "-carboxamide" if the carbon in the amide group cannot be included in the main chain. The prefix form is both "carbamoyl-" and "amido-".

Amides that have additional substituents on the nitrogen are treated similarly to the case of amines: they are ordered alphabetically with the location prefix *N*: $HCON(CH_3)_2$ is *N*,*N*-dimethylmethanamide.

Cyclic Compounds

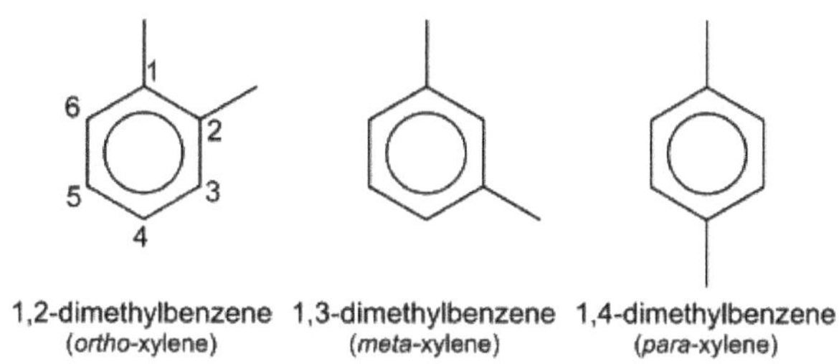

Cycloalkanes and aromatic compounds can be treated as the main parent chain of the compound, in which case the positions of substituents are numbered around the ring structure. For example, the three isomers of xylene $CH_3C_6H_4CH_3$, commonly the *ortho-*, *meta-*, and *para-* forms, are 1,2-dimethylbenzene, 1,3-dimethylbenzene, and 1,4-dimethylbenzene. The cyclic structures can also be treated as functional groups themselves, in which case they take the prefix cyclo*alkyl-*" (e.g. "cyclohexyl-") or for benzene, "phenyl-".

The IUPAC nomenclature scheme becomes rapidly more elaborate for more complex cyclic structures, with notation for compounds containing conjoined rings, and many common names such as phenol being accepted as base names for compounds derived from them.

Spiro Compounds

Bicyclic Compounds

Polycyclic Compounds

Order of Precedence of Groups

When compounds contain more than one functional group, the order of precedence determines which groups are named with prefix or suffix forms. The highest-precedence group takes the suffix, with all others taking the prefix form. However, double and triple bonds only take suffix form (-en and -yn) and are used with other suffixes.

Prefixed substituents are ordered alphabetically (excluding any modifiers such as di-, tri-, etc.), e.g. chlorofluoromethane, *not* fluorochloromethane. If there are multiple functional groups of the same type, either prefixed or suffixed, the position numbers are ordered numerically (thus ethane-1,2-diol, *not* ethane-2,1-diol.) The *N* position indicator for amines and amides comes before "1", e.g. $CH_3CH(CH_3)CH_2NH(CH_3)$ is *N*,2-dimethylpropanamine.

Priority	Functional group	Formula	Prefix	Suffix
1	Cations		-onio-	-onium
	e.g. Ammonium	NH_4^+	ammonio-	-ammonium
2	Carboxylic acids	$-COOH$	carboxy-	-oic acid*
	Carbothioic *S*-acids	$-COSH$	sulfanylcarbonyl-	-thioic *S*-acid*
	Carboselenoic *Se*-acids	$-COSeH$	selanylcarbonyl-	-selenoic *Se*-acid*
	Sulfonic acids	$-SO_3H$	sulfo-	-sulfonic acid
	Sulfinic acids	$-SO_2H$	sulfino-	-sulfinic acid
3	*Carboxylic acid derivatives*			
	Esters	$-COOR$	R-oxycarbonyl-	-R-oate
	Acyl halides	$-COX$	halocarbonyl-	-oyl halide*
	Amides	$-CONH_2$	carbamoyl-	-amide*
	Imides	$-CON{=}C<$	-imido-	-imide*
	Amidines	$-C(=NH)NH_2$	amidino-	-amidine*
4	Nitriles	$-CN$	cyano-	-nitrile*
	Isocyanides	$-NC$	isocyano-	isocyanide
5	Aldehydes	$-CHO$	formyl-	-al*
	Thioaldehydes	$-CHS$	thioformyl-	-thial*

6	Ketones Thiones Selones Tellones	=O =S =Se =Te	oxo- sulfanylidene- selanylidene- tellanylidene-	-one -thione -selone -tellone
7	Alcohols Thiols Selenols Tellurols	–OH –SH –SeH –TeH	hydroxy- sulfanyl- selanyl- tellanyl-	-ol -thiol -selenol -tellurol
8	Hydroperoxides Peroxols Thioperoxols (Sulfenic acid) Dithioperoxols	-OOH -SOH -SSH	hydroperoxy- hydroxysulfanyl- disulfanyl-	-peroxol -*SO*-thioperoxol -dithioperoxol
9	Amines Imines Hydrazines	$–NH_2$ =NH $–NHNH_2$	amino- imino- hydrazino-	-amine -imine -hydrazine

**Note*: These suffixes, in which the carbon atom is counted as part of the preceding chain, are the most commonly used. See individual functional group articles for more details.

The order of remaining functional groups is only needed for substituted benzene and hence is not mentioned here

COMMON NOMENCLATURE – TRIVIAL NAMES

Common nomenclature uses the older names for some organic compounds instead of using the prefixes for the carbon skeleton above. The pattern can be seen below.

Number of carbons	Prefix as in new system	Common name for alcohol	Common name for aldehyde	Common name for acid
1	Meth-	**Meth**yl alcohol (wood alcohol)	**Form**aldehyde	**Form**ic acid
2	Eth-	**Eth**yl alcohol (grain alcohol)	**Acet**aldehyde	**Acet**ic acid
3	Prop-	**Prop**yl alcohol	**Propion**aldehyde	**Propion**ic acid
4	But-	**But**yl alcohol	**Butyr**aldehyde	**Butyr**ic acid
5	Pent-	**Am**yl alcohol	**Valer**aldehyde	**Valer**ic acid
6	Hex-	**Capro**yl alcohol	**Capro**aldehyde	**Capro**ic acid
7	Hept-	**Enanth**yl alcohol	**Enanth**aldehyde	**Enanth**oic acid
8	Oct-	**Capr**yl alcohol	**Capryl**aldehyde	**Capryl**ic acid

9	Non-	**Pelargon**ic alcohol	**Pelargon**aldehyde	**Pelargon**ic acid
10	Dec-	**Capr**ic alcohol	**Capr**aldehyde	**Capr**ic acid
11	Undec-	-	-	-
12	Dodec-	**Laur**yl alcohol	**Laur**aldehyde	**Laur**ic acid
13	Tridec-	-	-	-
14	Tetradec-	**Myrist**yl alcohol	**Myrist**aldehyde	**Myrist**ic acid
15	pentadec-			
16	Hexadec-	**Cet**yl alcohol **Palmit**yl alcohol	**Palmit**aldehyde	**Palmit**ic acid
17	Heptadec-	-	-	**Margar**ic acid
18	Octadec-	**Stear**yl alcohol	**Stear**aldehyde	**Stear**ic acid
20	Eicos-	**Arachid**yl alcohol	-	**Arachid**ic acid
22	Docos-	**Behen**yl alcohol	-	**Behen**ic acid
24	Tetracos-	**Lignocer**yl alcohol	-	**Lignocer**ic acid
26	Hexacos-	**Cer**yl alcohol	-	**Cerot**ic acid
28	Octacos-	**Montan**yl alcohol	-	**Montan**ic acid
30	Triacont-	**Meliss**yl alcohol	-	**Meliss**ic acid
32	Dotriacont-	**Laccer**yl alcohol	-	**Laccer**oic acid
33	Tritriacont-	**Psyll**ic alcohol	-	**Psyll**ic acid
34	Tetratriacont-	**Gedd**yl alcohol	-	**Gedd**ic acid
35	Pentatriacont-	-	-	**Ceroplast**ic acid
40	Tetracont-	-	-	-

Ketones

Common names for ketones can be derived by naming the two alkyl or aryl groups bonded to the carbonyl group as separate words followed by the word *ketone*.

- Acetone
- Acetophenone
- Benzophenone
- Ethyl isopropyl ketone
- Diethyl ketone

The first three of the names shown above are still considered to be acceptable IUPAC names.

Aldehydes

The common name for an aldehyde is derived from the common name of the corresponding carboxylic acid by dropping the word *acid* and changing the suffix from -ic or -oic to -aldehyde.

- Formaldehyde
- Acetaldehyde

IONS

The IUPAC nomenclature also provides rules for naming ions.

Hydron

Hydron is a generic term for hydrogen cation; protons, deuterons and tritons are all hydrons. The Hydrons are not found in heavier isotopes, however.

Parent Hydride Cations

Simple cations formed by adding a hydron to a hydride of a halogen, chalcogen or pnictogen are named by adding the suffix "-onium" to the element's root: H_4N^+ is ammonium, H_3O^+ is oxonium, and H_2F^+ is fluoronium. Ammonium was adopted instead of nitronium, which commonly refers to NO_2^+.

If the cationic center of the hydride is not a halogen, chalcogen or pnictogen then the suffix "-ium" is added to the name of the neutral hydride after dropping any final 'e'. H_5C^+ is methanium, HO-(O^+)-H_2 is dioxidanium (HO-OH is dioxidane), and H_2N-(N^+)-H_3 is diazanium (H_2N-NH_2 is diazane).

Cations and Substitution

The above cations except for methanium are not, strictly speaking, organic, since they do not contain carbon. However, many organic cations are obtained by substituting another element or some functional group for a hydrogen.

The name of each substitution is prefixed to the hydride cation name. If many substitutions by the same functional group occur, then the number is indicated by prefixing with "di-", "tri-" as with halogenation. $(CH_3)_3O^+$ is trimethyloxonium. $CH_3F_3N^+$ is trifluoromethylammonium.

POLYCYCLIC COMPOUND

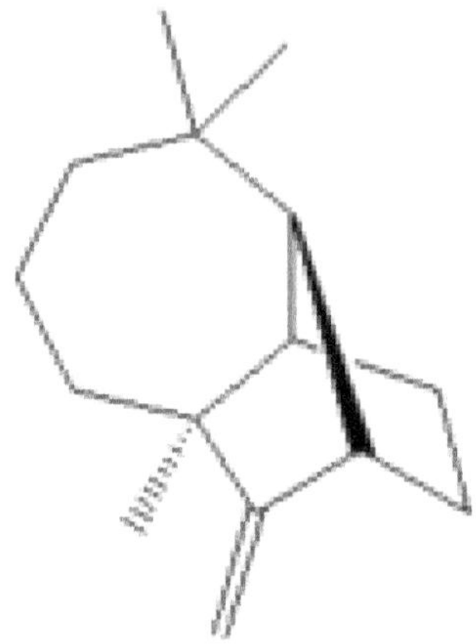

Longifolene, a terpene natural product, and an example of a tricyclic molecule.

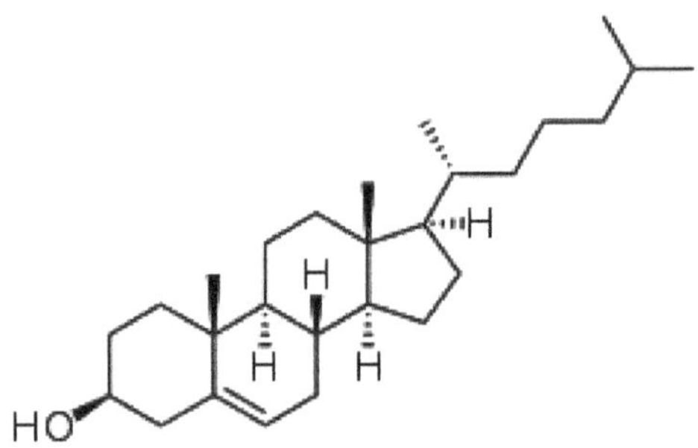

Cholesterol, another terpene natural product, in particular, a steroid, a class of tetracyclic molecules.

In the field of organic chemistry, a **polycyclic compound** is an organic chemical featuring several closed rings of atoms, primarily carbon. These ring substructures include cycloalkanes, aromatics-, and other ring types. They come in sizes of three atoms and upward, and in combinations of linkages that include tethering (such as in biaryls), fusing (edge-to-edge, such as in anthracene and steroids), links via a single atom (such as in spiro compounds), and bridged cyclics such as the longifolene example. Though poly- literally means "many", there is some latitude in determining how many rings are required to be considered polycyclic; many smaller rings are described by specific prefixes (e.g., bicyclic, tricyclic, tetracyclic, etc.), and so while it can refer to these, the title term is used with most specificity when these alternative names and prefixes are unavailable.

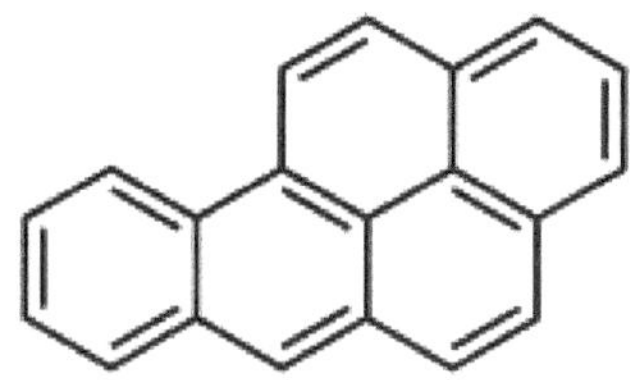

Benzo[a]pyrene, a pentacyclic compound both natural and man-made.

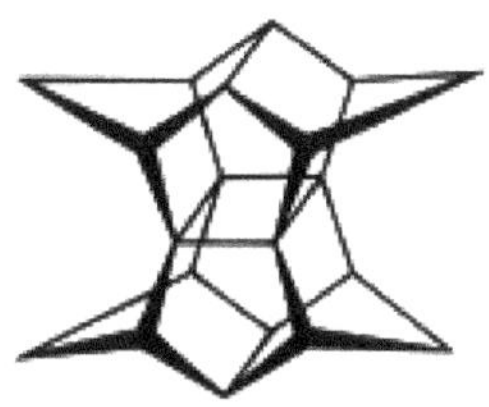

Pagodane, a man-made polycyclic compound.

In general, the term polycyclic includes polycyclic aromatic compounds, including polycyclic aromatic hydrocarbons, as well as heterocyclic aromatic compounds with multiple rings (where heteroaromatic compounds are aromatic compounds that contain sulfur, nitrogen, oxygen, or another non-carbon atoms in their rings in addition to carbon).

BICYCLIC MOLECULE

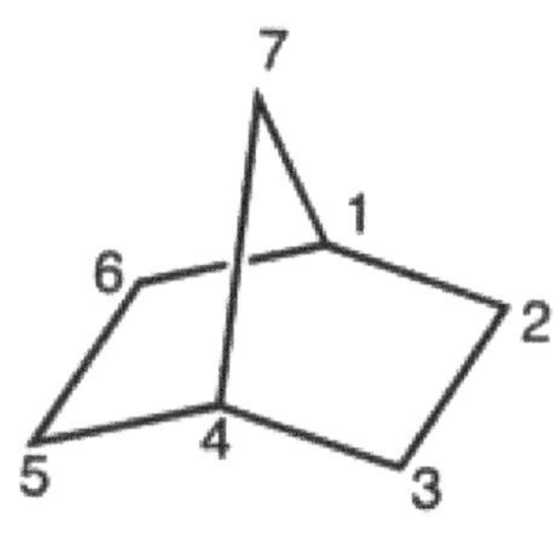

The bridged bicyclic norbornane, formally bicyclo[2.2.1]heptane.

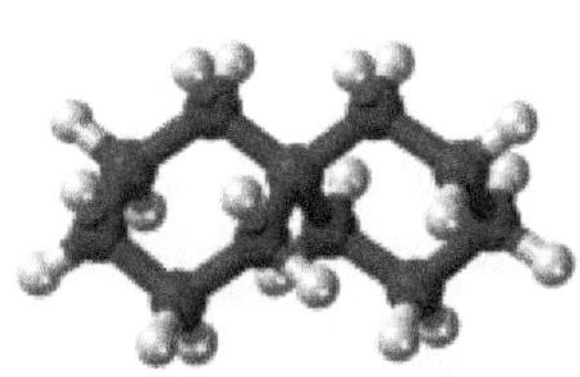

The spirocyclic compound spiro[5.5]undecane.

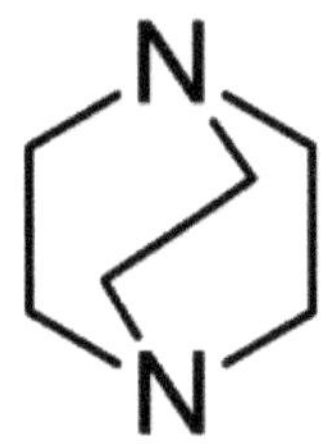

DABCO, a heterocyclic, bridged bicyclic compound, formally named 1,4-diazabicyclo[2.2.2] octane.

A **bicyclic molecule** (*bi* = two, *cycle* = ring) is a molecule that features two joined rings. Bicyclic structures occur widely, for example in many biologically important molecules like α-thujene and camphor. A bicyclic compound can be carbocyclic (all of the ring atoms are carbons), or heterocyclic (the rings atoms consist of at least two different elements), like DABCO. Moreover, the two rings can both be aliphatic (*e.g.* decalin and norbornane), or can be aromatic (*e.g.*naphthalene), or a combination of aliphatic and aromatic (*e.g.* tetralin).

There are three possible modes of ring junction for a bicyclic compound:

- In **spirocyclic compounds**, the two rings share only one single atom, the spiro atom, which is usually a quaternary carbon. An example of a spirocyclic compound is the photochromic switch spiropyran.
- In **fused bicyclic compounds**, two rings share two adjacent atoms. In other words, the rings share one covalent bond, *i.e.* the so-called bridgehead atoms are directly connected (*e.g.* α-thujene and decalin).
- In **bridged bicyclic compounds**, the two rings share three or more atoms, separating the two bridgehead atoms by a bridge containing at least one atom. For example, norbornane, also known as bicyclo[2.2.1]heptane, can be thought of as a pair of cyclopentane rings each sharing three of their five carbon atoms. The structure of camphor has the same basis as norbornane, but with some substituents added.

Nomenclature

Bicyclic molecules have a strict nomenclature. The root of the compound name depends on the total number of atoms in all rings together, possibly followed by a suffix denoting the functional group with the highest priority. Numbering of the carbon chain always begins at one bridgehead atom (where the rings meet) and follows the carbon chain along the longest path, to the next bridgehead atom. Then numbering is continued along the second longest path and so on. Fused and bridged bicyclic compounds get the prefix *bicyclo*, whereas spirocyclic compounds get the prefix *spiro*. In between the prefix and the suffix, a pair of brackets with numerals denotes the number of carbon atoms between each of the bridgehead atoms. These numbers are arranged in descending order and are separated by periods. For example, the carbon frame of norbornane contains a total of 7 atoms, hence the root name heptane. This molecule has two paths of 2 carbon atoms and a third path of 1 carbon atom between the two bridgehead carbons, so the brackets are filled in descending order: [2.2.1]. Addition of the prefix *bicyclo* gives the total name bicyclo[2.2.1]heptane.

The carbon frame of camphor also counts 7 atoms, but is substituted with a carbonyl in this case, hence the suffix heptanone. We start with numbering the carbon frame at the bridgehead atom with the highest priority (methyl goes before proton), hence the bridgehead carbon in front gets number 1, the carbonyl gets number 2 and numbering continues along the carbon chain following the longest path, until the doubly substituted top carbon (number 7). Equal to norbornane, this molecule also has two paths of 2 carbon atoms and one path of 1 carbon atom between the two bridgehead carbons, so the numbers within the brackets stay [2.2.1]. Combining the brackets and suffix (now filling in the position of the carbonyl as well) gives us [2.2.1]heptan-2-one. Besides *bicyclo*, the prefix should also specify the positions of all methyl substituents so the complete, official name becomes 1,7,7-trimethylbicyclo[2.2.1]heptan-2-one.

When naming simple fused bicyclic compounds, the same method as for bridged bicyclic compounds is applied, except the third path between the two bridgehead atoms now consists of zero atoms. Therefore, fused bicyclic compounds have a 0 included in the brackets. For example, decalin is named bicyclo[4.4.0]decane.

The heterocyclic molecule DABCO has a total of 8 atoms in its bridged structure, hence the root name octane. Note that here the two bridgehead atoms are nitrogen instead of carbon atoms. Therefore, the official name gets the additional prefix *1,4-diaza* and the total name becomes 1,4-diazabicyclo[2.2.2]octane.

SPIRO COMPOUND

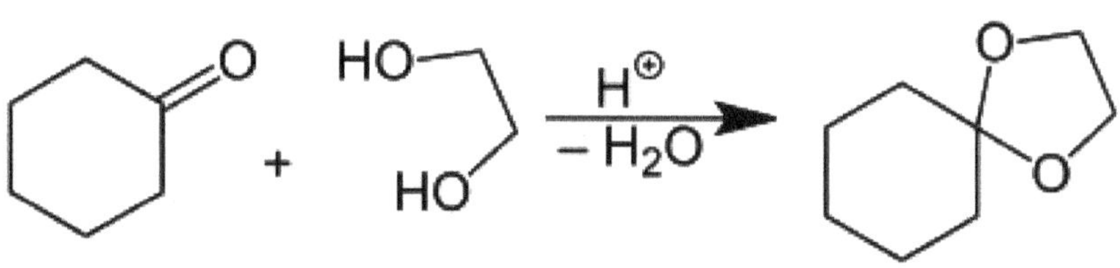

Preparation of a heterocyclic spiro compound. Each molecule shown in line-angle representation, where each angle (vertex) represents a carbon atom, and each line a bond between atoms (and where necessary hydrogen atoms to fulfill carbon's valence of four are implied). The heterocyclic spiro product, at right of arrow, is a common ketal (acetal) structure prepared from a monocyclic type of ketone and a generally acyclic type of diol (both at left of arrow). The schematic is for the synthesis of the protected 1,4-dioxaspiro[4.5]decane from cyclohexanone and the alcohol, ethanediol. Note the apparent planarity, which masks the actual orthogonal orientation of the two rings.

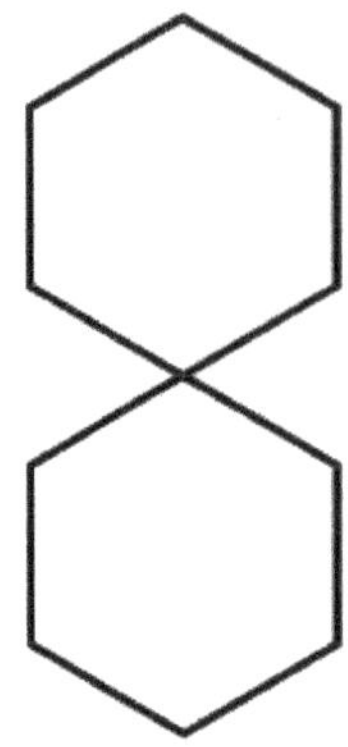

A fully carbo-cyclic spiro com-pound. Spiro[5.5]undecane,again shown in line-angle representation and in apparent planarity, masking the actual most populated chair conformation of each ring, and their actual orthogonal orientation to one another.

A **spiro compound**, or **spirane**, from the Latin *spīra*, meaning a twist or coil,[:1138] is a chemical compound, typically an organic compound, that presents a twisted structure of two or more rings (a ring system), in which 2 or 3 rings are linked together by one common atom,examples of which are shown at right.

The simplest spiro compounds are bicyclic (having just two rings), or have a bicyclic portion as part of the larger ring system, in either case with the two rings connected through the defining single common atom. The one common atom connecting the participating rings distinguishes spiro compounds from other bicyclics: from *isolated ring compounds* like biphenyl that have no connecting atoms, from *fused ring compounds* like decalin having two rings linked by two adjacent atoms, and from *bridged ring compounds* like norbornane with two rings linked by two non-adjacent atoms.

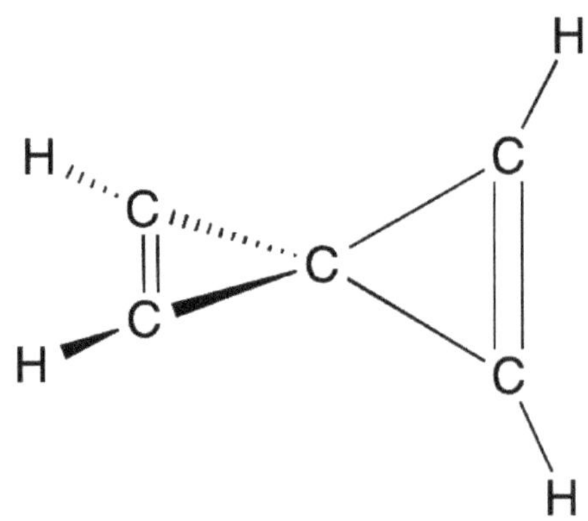

A fully carbocyclic spiro compound. Spiro-pentadiene, an esoteric hydrocarbon, in a representation where all carbons and hydrogens are shown explicitly. Note the two cyclopropene rings, each being planar, and the two being non-coplanar (perpendicular) to each other.

Spiro compounds may be fully carbocyclic (all carbon) or heterocyclic (having one or more non-carbon atom), and the earliest exposure of a chemist in training to a spiro compound is often to a heterocyclic form, the ketal (acetal) formed in the protection of ketones by diols and dithiols (see image above right). The common atom that connects the two (or sometimes three) rings is called the *spiro atom*; in carbocyclic spiro compounds like spiro[5.5]undecane (see image at right), the spiro-atom is a quaternary carbon, and as the -ane ending implies, these are the types of molecules to which the name *spirane* was first applied (though it is now used general of all spiro compounds).[:1138ff] Likewise, a tetravalent neutral silicon or positively charged quaternary nitrogen atom (ammonium cation) can be the spiro center in these compounds, and many of these have been prepared and described. The 2-3 rings being joined are most often different in nature, though they, on occasion, be identical [e.g., spiro[5.5]undecane, just shown, and spirapentadiene, at right]. Although sketches of organic structures makes spiro compounds appear planar, they are not; for instance, a spiro compound with a pair of three-membered cyclopropene rings connected in spiro fashion (image below) has been given the popular misnomer of being a bow tie structure, when it is not flat or planar like a bow tie. This can be stated another way, saying that the best-fit planes to each ring are often perpendicular or are otherwise non-coplanar to one another.

Spiranes can be chiral, in three distinct ways. First, while nevertheless appearing to be twisted, they yet may have a chiral center making them analogous to any simple chiral compound, and second, while again appearing twisted, the specific location of substiuents, as with alkylidenecycloalkanes, may make a spiro compound display *central chirality* (rather than axial chirality resulting from the twist); third, the substiuents of the rings of the spiro compound may be such that the only reason they are chiral arises solely from the twist of their rings, e.g., in the simplest bicyclic case, where two structurally identical rings are attached via their spiro atom, resulting in a twisted presentation of the two rings. Hence, in the third case, the lack of planarity described above gives rise to what is termed axial chirality in otherwise identical isomeric pair of spiro compounds, because they differ only in the right- *versus* left-handed "twist" of structurally identical rings (as seen in allenes, sterically hindered biaryls, and alkylidenecycloalkanes as well). Assignment of absolute configuration of spiro compounds has been challenging, but a number of each type have been unequivocally assigned.

Spiro compounds are present throughout the natural world, some cases of which have been exploited to provide tool compounds for biomedical study and to serve as scaffolds for the design of therapeutic agents with novel shapes. As well, the spiro motif is present in various practical compound types (such as dyes), as well as in a wide variety of oligo- and polymeric materials designs, for the unique shapes and properties the spiro center imparts, e.g., in the design of electronically active materials in particular. In both cases, the presence of the spiro center, often with four distinct groups attached, and with its unique aspects of chirality, adds unique challenges to the chemical synthesis of each compound type.

Carbocyclic Spiro Compounds

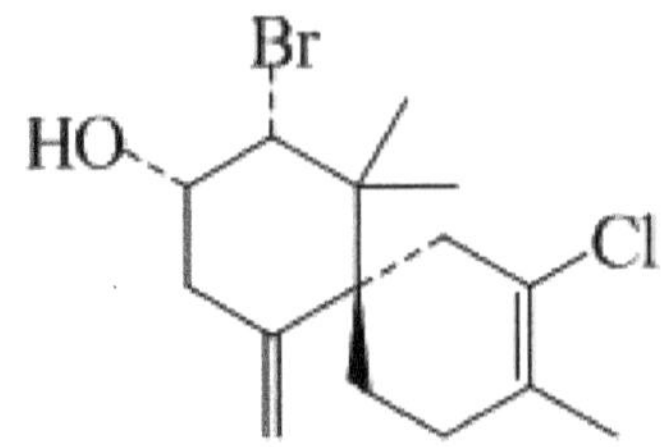

The Brazilian *Laurencia dendroidea* (red algae) sesquiterpene and halogenated carbocyclic spiro compound, elatol.

Bicyclic ring structures in organic chemistry that have two fully carbocyclic (all carbon) rings connected through just one atom are present both in natural products, as well as in esoteric targets of chemical synthesis. The two carbocycles can be different in nature, or identical. In common targets derived from natural products, they are essentially always different. In esoteric targets, such as highly strained hydrocarbons like spiropentadiene, shown here, the rings can be identical. The atom connecting the two rings is called the *spiro-atom*; in carbocyclic spiro compounds, the spiro-atom is a quaternary carbon. The 11-carbon bicyclic structure shown above, spiro[5.5]undecane, is also a fully carbocyclic spiro compound. While the presentation of this structure makes it appear fully planar, it is not. The best-fit planes to each 6-atom ring above is near to perpendicular, and the best-fit planes to rings of spiro compounds are likewise generally non-coplanar. For instance, the structure of faux bow tie spiropentadiene, shown above, makes clear that the planes that are defined by the there atoms of each ring—i.e., the best-fit plane of each cyclopropene—are orthogonal (perpendicular) to one another.

Heterocyclic Spiro Compounds

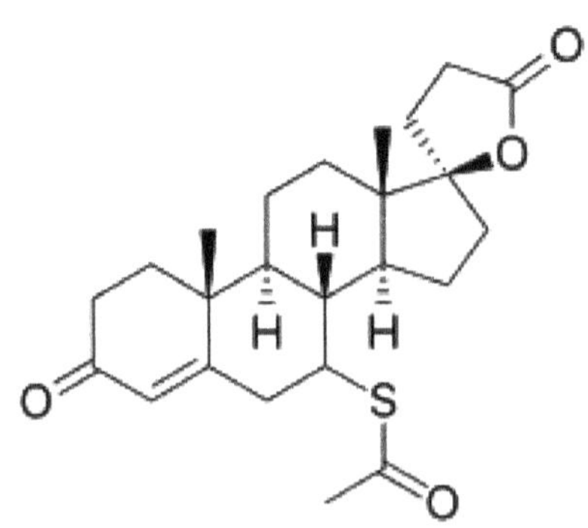

A further example of a hetero-cyclic spiro compound, the on-market drug spironolactone, where the two 5-membered rings ring in the upper right hand corner compose a spiro ring system. One of the rings is a lactone (i.e., contains as ester functional group).

Spiro compounds are considered heterocyclic if the spiro atom or any atom in either ring are not carbon atoms. Cases include the presence of a spiro heteroatom such silicon and nitrogen (but also other Group IVA and other atom types) connecting the rings that have been observed or are under theoretical study;[:1139f] moreover, there are also many cases where one or more heteroatoms appear in one or more of the rings that are joined at a carbon spiro atom (e.g., where 1 oxygen spironolactones and 2 oxygen/2 sulfur ketals/ thioketals are very common).

A common case is the presence of two atoms that are not carbon in one of the rings, with those two rings both attached to the spiro atom; indeed, often the earliest exposure of a chemist in training to a spiro compound is to a heterocyclic form, the ketal (acetal) formed in the protection of ketones by diols and dithiols. An example of this is shown above, in the synthesis of the acetal 1,4-dioxaspiro[4.5]decane from cyclohexanone and ethanediol. In this case, because the four atoms attached to the spiro atom are not all carbons, the spiro atom is not a quaternary carbon. A further example of an acetal formed from a cyclic ketone, except with a dithiol, is the spiro compound spirapril, which has a five-membered ring formed from 1,2-ethanedithiol. Again, while the rings could be identical, in the heterocyclic case they are, again, almost always non-identical. Once again, the best-fit planes to each ring are generally non-coplanar to one another (i.e., the rings are not coplanar, despite appearing so in images).

Polyspiro Compounds

A polyspiro compound is connected by two or more spiroatoms making up three or more rings.

Nomenclature

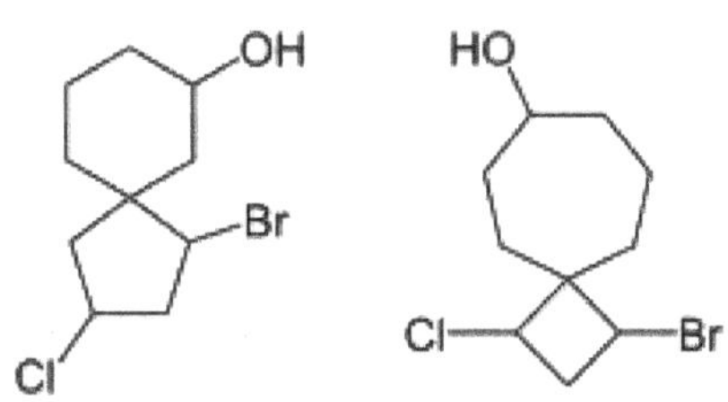

A B

Two examples of spiro compound nomenclature, **A.** 1-bromo-3-chloro-spiro[4.5]-decan-7-ol, and **B.** 1-bromo-3-chloro-spiro[3.6]-decan-7-ol.

Nomenclature for spiro compounds was first discussed by Adolf von Baeyer in 1900. Spiro compounds are named with the infix *spiro* followed by square brackets containing the number of atoms in the smaller ring, then the number of atoms in the larger ring, separated by a period, in each case excluding the spiroatom itself. For example, compound A is called *1-bromo-3-chlorospiro[4.5]decan-7-ol*, and compound B is called *1-bromo-3-chlorospiro[3.6]decan-7-ol*.

Chirality

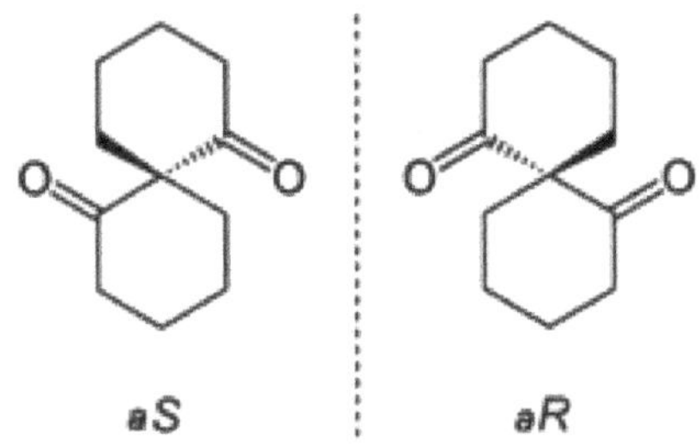

Some spiro compounds exhibit axial chirality. Spiroatoms can be the origin of chirality even when they lack the required four different substituents normally observed in chirality. When two rings are identical the priority is determined by a slight modification of the CIP system assigning a higher priority to one ring extension and a lower priority to an extension in the other ring. When rings are dissimilar the regular rules apply.

Preparation

Spiro compounds present unique preparative challenges, whether each ring contributing to its structure is unique or identical, or whether they are carbocyclic or heterocyclic—owing to the practical implications of tetra-functiionalizing the central spiro atom (often with four different groups), and of the unique aspects of chirality that apply to these compounds.

Specific Methods

Some spiro compounds can be synthesized using the Pinacol-pinacolone rearrangement;[:985] for example, spiro[5.4]decane (final compound in following two line scheme) can be synthesized from symmetric 1,2-diols of the sort shown below [e.g., this route›s starting material, (1,1′-bicyclopentyl)-1,1′-diol]. Initially, one of the carbinol moieties is protonated, allowing water to leave, and yielding the corresponding carbocation (second structure, first row); this intermediate then undergoes a bond migration, resulting in ring expansion of the adjacent ring, with deprotionation unmasking the ketone functional group to complete the first line of the mechanism. This first product, a spirobicyclic ketone, is a spiro compound in its own right, and yields the further spiro carbinol and the alicyclic spiro hydrocarbon after two further reduction reactions. First, reduction of the carbonyl that ends the mechanism›s first line provides the spiro carbinol starting material of the second line, which is needed for reduction to the alkane (shown). This latter reduction is accomplished using lithium aluminum hydride ($LiAlH_4$), via the alcohol tosylate (formed using tosyl chloride). Hence this three reaction sequence provides three spiro compounds (ketone, alcohol, and alkane), of possible research or practical use.[:985]

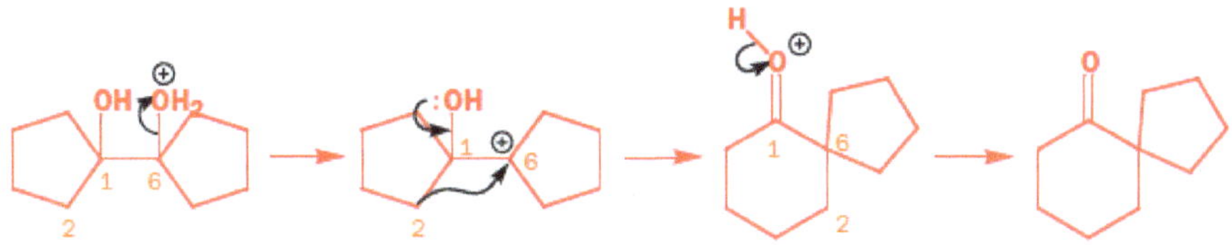

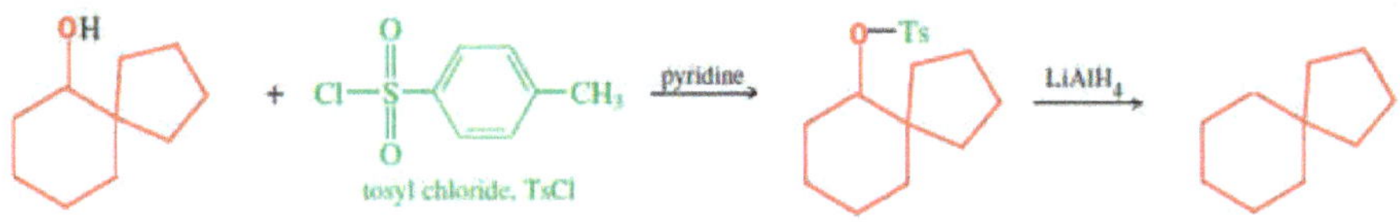

REFERENCES

1. 1,4-Dioxaspiro[4.5]decane . chemspider.com. Retrieved 3 February 2016.
2. spiro[5.5]undecane . chemspider.com. Retrieved 3 February 2016.
3. Eliel, et al., op. cit., introduces the synonym spirane and the Latin and translation as twist or whorl; Lewis› dictionary, op. cit., speaking to basic definitions in ancient use, and provides the vowel marking and definitions of coil, fold, twist, or spiral.
4. Eliel, Ernest Ludwig; Wilen, Samuel H. & Mander, Lewis N. (1994). Chirality in Molecules Devoid of Chiral Centers (Chapter 14) . Stereochemistry of Organic Compounds (1st ed.). New York, NY, USA: Wiley & Sons. pp. 1119–1190, esp. 1119ff, 1138ff; and passim. ISBN 0471016705. Retrieved 2 February 2016. For a further but less stable source of the same text that provides access to the relevant material, see , same access date.
5. Lewis, Charlton, T. (1890). spīra [dictionary entry] . An Elementary Latin Dictionary. New York, NY, USA: American Book Company. Retrieved 3 February 2016. Quoting: ‹spīra ae, f, *σπερα, a coil, fold, twist, spiral: in* spirain se conligit anguis, V., O.: longo iactetur spira galero, i. e. tie, Iu.› The Greek transcription, σπερα, reflects the use of this cognate as one ancient Greek term to refer to a coil or related fold, see Woodhouse, S.C. (1910). Fold, subs. [dictionary entry] . English-Greek Dictionary: A Vocabulary of the Attic Language. Ludgate Hill [London, ENG]: George Routledge & Sons. Retrieved 3 February 2016. Quoting: ‹Fold, subs. ... Coil : *V. σϖερα... see coil.'*
6. Moss, G.P. and the Working Party of the International Union of Pure and Applied Chemistry [IUPAC], Organic Chemistry Division, Commission on Nomenclature of Organic Chemistry (III.1) (1999). Extension and Revision of the Nomenclature for Spiro Compounds (IUPAC Recommendations1999) (PDF). Pure Appl. Chem. **71** *(3): 531–558.* doi:10.1351/pac199971030531. ISSN 1365-3075. Retrieved 3 February 2016. Note, the article co-authors, the Working Party of the IUPAC (1992-1998), were P. M. Giles, Jr., E. W. Godly, K.-H. Hellwich, A. K. Ikizler, M. V. Kisakürek, A. D. McNaught, G. P. Moss, J. Nyitrai, W. H. Powell, O. Weissbach, and A. Yerin. Also available online at Extension and Revision of the Nomenclature

for Spiro Compounds . London, GBR: Queen Mary University of London. Retrieved 3 February 2016. Also available in German, with et al. indicating the same working party, atHellwich, Karl-Heinz et al. (2002). Erweiterung und Revision der Nomenklatur der Spiroverbindungen . Angewandte Chemie. **114** *(20, 18 October): 4073–4089.* doi:10.1002/1521-3757(20021018)114:20<4073::AID-ANGE4073>3.0.CO;2-T. Retrieved 3 February 2016. Die *Übersetzung* basiert auf der „Extension and Revision of the Nomenclature for Spiro Compounds“ der Commission on Nomenclature of Organic Chemistry (III.1) der Organic Chemistry Division der International Union of Pure and Applied Chemistry, veröffentlicht in Pure Appl. Chem.1999, 71, 531–558.

7. Clayden, Jonathan ; Greeves, Nick & Warren, Stuart (2012). Organic Chemistry (2nd ed.). Oxford, UK: Oxford University Press. pp. 319f, 432, 604, 653, 746, 803, 839, 846f. ISBN 0199270295. Retrieved 2 February 2016.
8. For all four categories, see Reusch, William (1999). Saturated Hydrocarbons, Alkanes and Cycloalkanes: Cycloalkanes (Table: Examples of Isomeric C_8H_{14} Bicycloalkanes) or Nomenclature: Cycloalkanes (same Table), and passim . Virtual Text of Organic Chemistry (Jan. 2016 ed.). East Lansing, MI, USA: Michigan State University, Department of Chemistry. Retrieved 3 February 2016.The specific chapters can be found at and , respectively, same access date. For the description featuring adjacent atoms for all but the isolated category, see Clayden, op. cit.
9. Smith, Laura K. & Baxendale, Ian R. (2015). Total Syntheses of Natural Products Containing Spirocarbocycles (online, print review). Org. Biomol. Chem. **13**: 9907–9933. doi:10.1039/C5OB01524C. Retrieved 2 February 2016.
10. Science News Staff (1991). Elusive bowtie pinned down (online). Science News. Washington, DC, USA: Science Service, Society for Science and the Public, Gale-Thomson (July 13). Retrieved 2 February 2016. Quoting: ‹W.E. Billups and Michael M. Haley of Rice University in Houston accepted the challenge… in the June 19 Journal of the American Chemical Society, they report producing the molecule… / The molecule, which resembles a bowtie with one side bent perpendicular to the other, belongs to a chemical family called the spiroalkenes, whose members possess a single carbon at their cores.›
11. Rios, Ramon (2012). Enantioselective Methodologies for the Synthesis of Spiro Compounds . Chem. Soc. Rev. (print, online). **41**: 1060–1074. doi:10.1039/C1CS15156H.
12. von Baeyer, Adolf (1900). Systematik und Nomenclatur Bicyclischer Kohlenwasserstoffe . Berichte der deutschen chemischen Gesellschaft. **33** *(3): 3771–3775. doi:10.1002/cber.190003303187.*
13. Pubchem. 1,1›-Bicyclopentyl-1,1›-diol . nih.gov. Retrieved 7 March 2016.

CHAPTER 4

Theory of Organic Reactions

Chapter Contents

- **Introduction**
- **Classifications**
- **Fundamentals**
- **Organic Reactions by Mechanism**
- **Organic Reactions by Functional Groups**
- **Other Organic Reaction Classification**
- **Adams's Catalyst**
- **Claisen Condensation**
- **Fatty Acid Metabolism**
- **Aldol Condensation**

An organic chemical, also referred to as an organic compound, is defined as being constructed of molecules that possess carbon-based atoms. Specifically, it is composed of carbon atoms attached to hydrogen atoms, together with a variety of other elements from functional groups, such as oxygen, nitrogen, sulfur, etc. Usually, an organic chemical is covalently bonded, meaning that electrons are shared between atoms. This characteristic is why organic chemicals can be structured in a variety of long chains and rings and form bonds with the great stability.

INTRODUCTION

Organic reactions are chemical reactions involving organic compounds. The basic organic chemistry reaction types are addition reactions, elimination reactions, substitution reactions, pericyclic reactions, rearrangement reactions, photochemical reactions and redox reactions. In organic synthesis, organic reactions are used in the construction of new organic molecules. The production of many man-made chemicals such as drugs, plastics, food additives, fabrics depend on organic reactions.

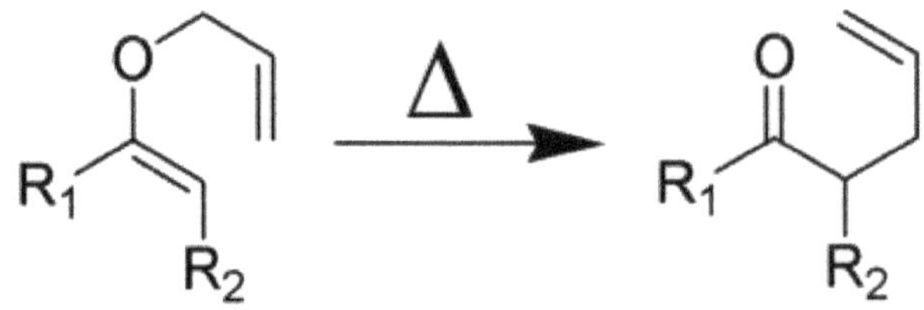

The oldest organic reactions are combustion of organic fuels and saponification of fats to make soap. Modern organic chemistry starts with the Wöhler synthesis in 1828. In the history of the Nobel Prize in Chemistry awards have been given for the invention of specific organic reactions such as the Grignard reaction in 1912, the Diels-Alder reaction in 1950, the Wittig reaction in 1979 and olefin metathesis in 2005.

CLASSIFICATIONS

Organic chemistry has a strong tradition of naming a specific reaction to its inventor or inventors and a long **list** of so-called **named reactions** exists, conservatively estimated at 1000. A very old named reaction is the Claisen rearrangement (1912) and a recent named reaction is the Bingel reaction (1993). When the named reaction is difficult to pronounce or very long as in the Corey-House-Posner-Whitesides reaction it helps to use the abbreviation as in the CBS reduction. The number of reactions hinting at the actual process taking place is much smaller, for example the ene reaction or aldol reaction.

Another approach to organic reactions is by type of organic reagent, many of them inorganic, required in a specific transformation. The major types are oxidizing agents such as osmium tetroxide, reducing agents such as Lithium aluminium hydride, bases such as lithium diisopropylamide and acids such as sulfuric acid.

FUNDAMENTALS

Factors governing organic reactions are essentially the same as that of any chemical reaction. Factors specific to organic reactions are those that determine the stability of reactants and products such as conjugation, hyperconjugation and aromaticity and the presence and stability of reactive intermediates such as free radicals, carbocations and carbanions.

An organic compound may consist of many isomers. Selectivity in terms of regioselectivity, diastereoselectivity and enantioselectivity is therefore an important criterion for many organic reactions. The stereochemistry of pericyclic reactions is governed by the Woodward–Hoffmann rules and that of many elimination reactions by the Zaitsev›s rule.

Organic reactions are important in the production of pharmaceuticals. In a 2006 review it was estimated that 20% of chemical conversions involved alkylations on nitrogen and oxygen atoms, another 20% involved placement and removal of protective groups, 11% involved formation of new carbon-carbon bond and 10% involved functional group interconversions.

ORGANIC REACTIONS BY MECHANISM

There is no limit to the number of possible organic reactions and mechanisms. However, certain general patterns are observed that can be used to describe many common or useful reactions. Each reaction has a stepwise reaction mechanism that explains how it happens, although this detailed description of steps is not always clear from a list of reactants alone. Organic reactions can be organized into several basic types. Some reactions fit into more than one category. For example, some substitution reactions follow an addition-elimination pathway. This overview isn't intended to include every single organic reaction. Rather, it is intended to cover the basic reactions.

Reaction type	Subtype	Comment
Addition reactions	electrophilic addition nucleophilic addition radical addition	include such reactions as halogenation, hydrohalogenation and hydration.
Elimination reaction		include processes such as dehydration and are found to follow an E1, E2 or E1cB reaction mechanism
Substitution reactions	nucleophilic aliphatic substitution	with S_N1, S_N2 and S_Ni reaction mechanisms
	nucleophilic aromatic substitution	
	nucleophilic acyl substitution	
	electrophilic substitution	
	electrophilic aromatic substitution	
	radical substitution	
Organic redox reactions		are redox reactions specific to organic compounds and are very common.
Rearrangement reactions	1,2-rearrangements pericyclic reactions metathesis	

In condensation reactions a small molecule, usually water, is split off when two reactants combine in a chemical reaction. The opposite reaction, when water is consumed in a reaction, is called hydrolysis. Many Polymerization reactions are derived from organic reactions. They are divided into addition polymerizations and step-growth polymerizations.

In general the stepwise progression of reaction mechanisms can be represented using arrow pushing techniques in which curved arrows are used to track the movement of electrons as starting materials transition to intermediates and products.

ORGANIC REACTIONS BY FUNCTIONAL GROUPS

Organic reactions can be categorized based on the type of functional group involved in the reaction as a reactant and the functional group that is formed as a result of this reaction. For example, in the Fries rearrangement the reactant is an ester and the reaction product an alcohol.

An overview of functional groups with their preparation and reactivity is presented below:

Functional group	Preparation	Reactions
Acid anhydride	preparation	reactions
Acyl halides	preparation	reactions
Acyloins	preparation	reactions
Alcohols	preparation	reactions
Aldehydes	preparation	reactions
Alkanes	preparation	reactions
Alkenes	preparation	reactions
Alkyl halides	preparation	reactions
Alkyl nitrites	preparation	reactions
Alkynes	preparation	reactions
Amides	preparation	reactions
Amine oxide	preparation	reactions
Amines	preparation	reactions
Arene compounds	preparation	reactions
Azides	preparation	reactions
Aziridines	preparation	reactions
Carboxylic acids	preparation	reactions
Cyclopropanes	preparation	reactions
Diazo compounds	preparation	reactions
Diols	preparation	reactions
Esters	preparation	reactions
Ethers	preparation	reactions
Epoxide	preparation	reactions
Haloketones	preparation	reactions
Imines	preparation	reactions
Isocyanates	preparation	reactions
Ketones	preparation	reactions
Lactams	preparation	reactions
Nitriles	preparation	reactions
Nitro compounds	preparation	reactions
Phenols	preparation	reactions
Thiols	preparation	reactions

OTHER ORGANIC REACTION CLASSIFICATION

In heterocyclic chemistry, organic reactions are classified by the type of heterocycle formed with respect to ring-size and type of heteroatom. See for instance the chemistry of indoles. Reactions are also categorized by the change in the carbon framework. Examples are ring expansion and ring contraction, homologation reactions, polymerization reactions, insertion reactions, ring-opening reactions and ring-closing reactions.

Organic reactions can also be classified by the type of bond to carbon with respect to the element involved. More reactions are found in organosilicon chemistry, organosulfur chemistry, organophosphorus chemistry and organofluorine chemistry. With the introduction of carbon-metal bonds the field crosses over to organometallic chemistry.

CH		CB															He
CLi	CBe	CC															
CNa	CMg	CN															
		CO															
		CF															
		Ne															
		CAl															
		CSi															
		CP															
		CS															
		CCl															
		CAr															
CK	CCa	CSc	CTi	CV	CCr	CMn	CFe	CCo	CNi	CCu	CZn	CGa	CGe	CAs	CSe	CBr	CKr
CRb	CSr	CY	CZr	CNb	CMo	CTc	CRu	CRh	CPd	CAg	CCd	CIn	CSn	CSb	CTe	CI	CXe
CCs	CBa		CHf	CTa	CW	CRe	COs	CIr	CPt	CAu	CHg	CTl	CPb	CBi	CPo	CAt	Rn
Fr	CRa		Rf	Db	CSg	Bh	Hs	Mt	Ds	Rg	Cn	Nh	Fl	Mc	Lv	Ts	Og
		↓															
		CLa	CCe	CPr	CNd	CPm	CSm	CEu	CGd	CTb	CDy	CHo	CEr	CTm	CYb	CLu	
		Ac	CTh	CPa	CU	CNp	CPu	CAm	CCm	CBk	CCf	CEs	Fm	Md	No	Lr	

Chemical bonds to carbon	
Core organic chemistry	Many uses in chemistry
Academic research, but no widespread use	Bond unknown

ADAMS'S CATALYST

Adams's catalyst, also known as **platinum dioxide**, is usually represented as platinum(IV) oxide hydrate, $PtO_2•H_2O$. It is a catalyst for hydrogenation and hydrogenolysis in organic synthesis. This dark brown powder is commercially available. The oxide itself is not an active catalyst, but it becomes active after exposure to hydrogen whereupon it converts to platinum black, which is responsible for reactions.

Preparation

Adams's catalyst is prepared from chloroplatinic acid H_2PtCl_6 or ammonium chloroplatinate, $(NH_4)_2PtCl_6$, by fusion with sodium nitrate. The first published preparation was reported by V. Voorhees and Roger Adams. The procedure involves first preparing a platinum nitrate which is then heated to expel nitrogen oxides.

$H_2PtCl_6 + 6\ NaNO_3 \rightarrow Pt(NO_3)_4 + 6\ NaCl_{(aq)} + 2\ HNO_3$

$Pt(NO_3)_4 \rightarrow PtO_2 + 4\ NO_2 + O_2$

The resulting brown cake is washed with water to free it from nitrates. The catalyst can either be used as is or dried and stored in a desiccator for later use. Platinum can be recovered from spent catalyst by conversion to ammonium chloroplatinate using aqua regia followed by ammonia.

Adams›s catalyst is used for many applications. It has shown to be valuable for hydrogenation, hydrogenolysis, dehydrogenation, and oxidation reactions. During the reaction, platinum metal (platinum black) is formed which has been cited to be the active catalyst. Hydrogenation occurs with syn stereochemistry when used on an alkyne resulting in a cis-alkene. Some of the most important transformations include the hydrogenation of ketones to alcohols or ethers (the latter product forming in the presence of alcohols and acids) and the reduction of nitro compounds to amines. However, reductions of alkenes can be performed with Adam's catalyst in the presence of nitro groups without reducing the nitro group. When reducing nitro compounds to amines, platinum catalysts are preferred over palladium catalysts to minimize hydrogenolysis. The catalyst is also used for the hydrogenolysis of phenyl phosphate esters, a reaction that does not occur with palladium catalysts. The pH of the solvent significantly affects the reaction course, and reactions of the catalyst are often enhanced by conducting the reduction in neat acetic acid, or solutions of acetic acid in other solvents.

Development

Prior to the development of Adams's catalyst, organic reductions were carried out using colloidal platinum or platinum black. The colloidal catalysts were more active but posed difficulties in isolating reaction products. This led to more widespread use of platinum black. In Adams's own words:

"...Several of the problems I assigned my students involved catalytic reduction. For this purpose we were using as a catalyst platinum black made by the generally accepted best method known at the time. The students had much trouble with the catalyst they obtained in that frequently it proved to be inactive even though prepared by the same detailed procedure which resulted occasionally in an active product. I therefore initiated a research to find conditions for preparing this catalyst with uniform activity."

CLAISEN CONDENSATION

The **Claisen condensation** is a carbon–carbon bond forming reaction that occurs between two esters or one ester and another carbonyl compound in the presence of a strong base, resulting in a β-keto ester or a β-diketone. It is named after Rainer Ludwig Claisen, who first published his work on the reaction in 1887.

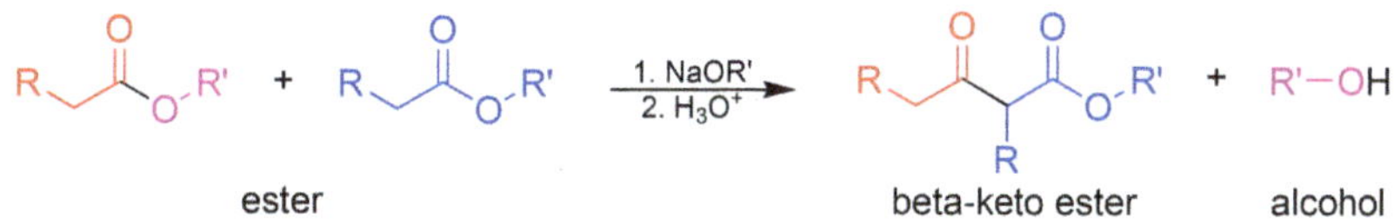

Requirements

At least one of the reagents must be enolizable (have an α-proton and be able to undergo deprotonation to form the enolate anion). There are a number of different combinations of enolizable and nonenolizable carbonyl compounds that form a few different types of Claisen condensations.

The base used must not interfere with the reaction by undergoing nucleophilic substitution or addition with a carbonyl carbon. For this reason, the conjugate sodium alkoxide base of the alcohol formed (e.g. sodium ethoxide if ethanol is formed) is often used, since the alkoxide is regenerated. In mixed Claisen condensations, a non-nucleophilic base such as lithium diisopropylamide, or LDA, may be used, since only one compound is enolizable. LDA is not commonly used in the classic Claisen or Dieckmann condensations due to enolization of the electrophilic ester.

The alkoxy portion of the ester must be a relatively good leaving group. Methyl and ethyl esters, which yields methoxide and ethoxide, respectively, are commonly used.

Types

- The classic Claisen condensation, a self-condensation between two molecules of a compound containing an enolizable ester.

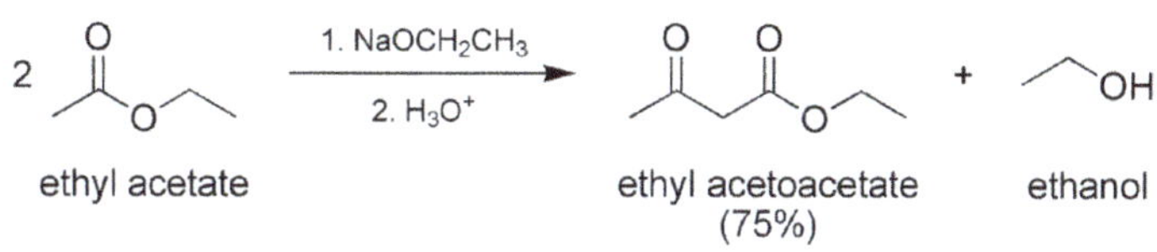

- The mixed (or "crossed") Claisen condensation, where one enolizable ester or ketone and one nonenolizable ester are used.

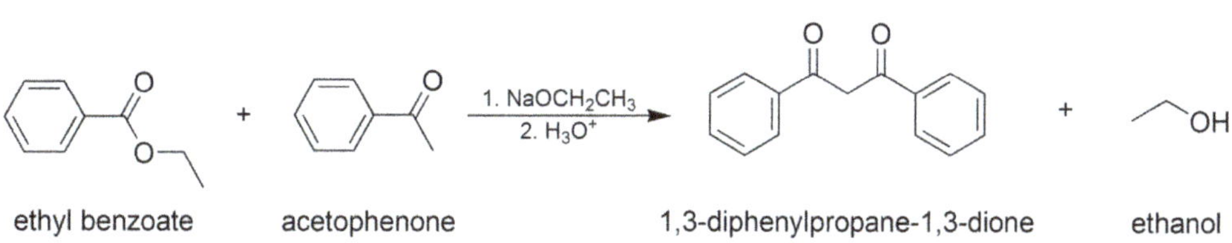

- The Dieckmann condensation, where a molecule with two ester groups reacts intramolecularly, forming a cyclic β-keto ester. In this case, the ring formed must not be strained, usually a 5- or 6-membered chain or ring.

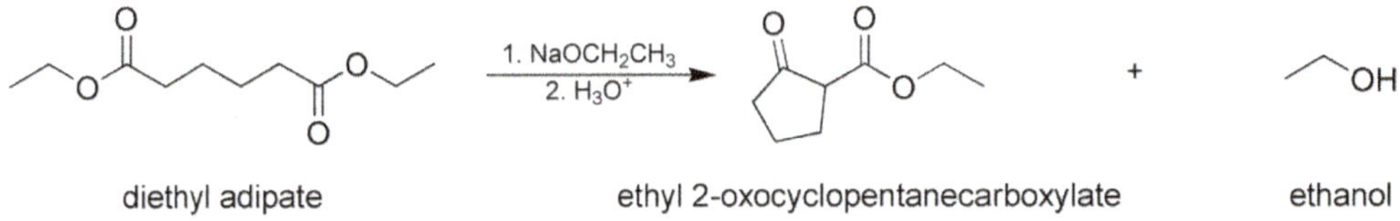

Mechanism

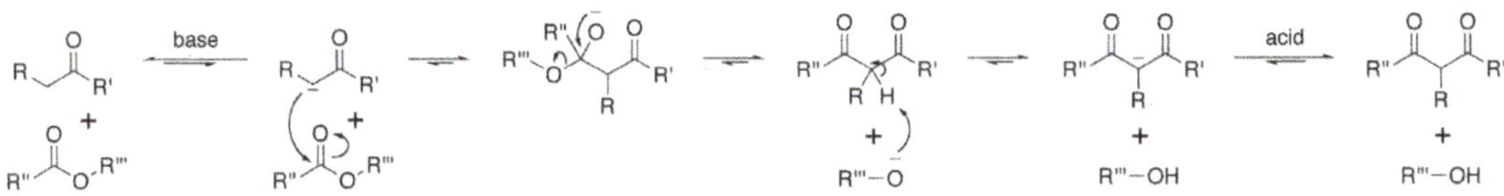

In the first step of the mechanism, an α-proton is removed by a strong base, resulting in the formation of an enolate anion, which is made relatively stable by the delocalization of electrons. Next, the carbonyl carbon of the (other) ester is nucleophilically attacked by the enolate anion. The alkoxy group is then eliminated (resulting in (re)generation of the alkoxide), and the alkoxide removes the newly formed doubly α-proton to form a new, highly resonance-stabilized enolate anion. Aqueous acid (e.g. sulfuric acid or phosphoric acid) is added in the final step to neutralize the enolate and any base still present. The newly formed β-keto ester or β-diketone is then isolated. Note that the reaction requires a stoichiometric amount of base as the removal of the doubly α-proton thermodynamically drives the otherwise endergonic reaction. That is, Claisen condensation does not work with substrates having only one α-hydrogen because of the driving force effect of deprotonation of the β-keto ester in the last step.

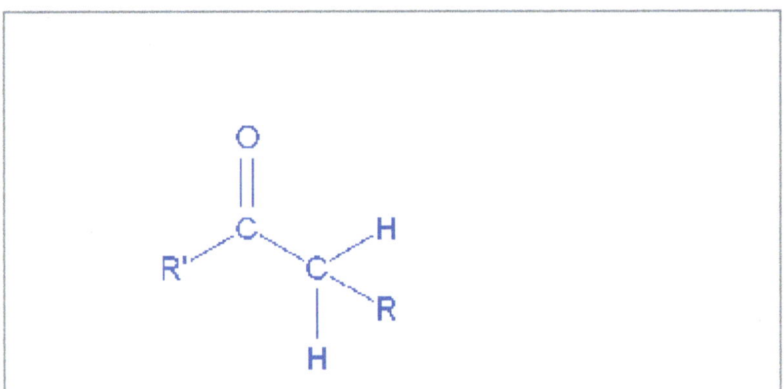

animation

Stobbe Condensation

The **Stobbe condensation** is a modification specific for the diethyl ester of succinic acid requiring less strong bases. An example is its reaction with benzophenone:

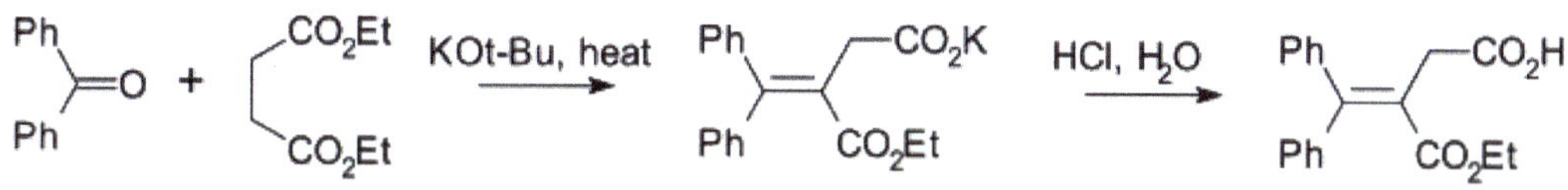

A reaction mechanism that explains the formation of both an ester group and a carboxylic acid group is centered on a lactone intermediate (**5**):

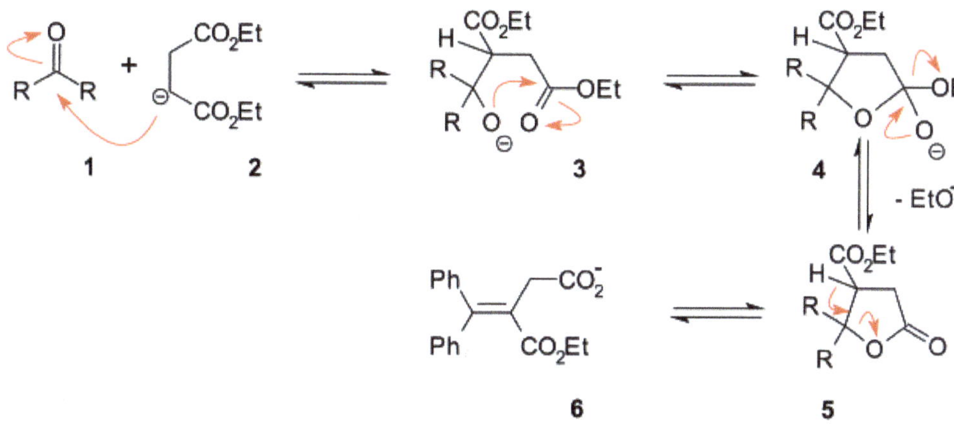

FATTY ACID METABOLISM

Fatty acid metabolism consists of catabolic processes that generate energy, and anabolic processes that create biologically important molecules (triglycerides, phospholipids, second messengers, local hormones and ketone bodies). Fatty acids are a family of molecules classified within the lipid macronutrient class. One role of fatty acids within animal metabolism is energy production, captured in the form of adenosine triphosphate (ATP). When compared to other macronutrient classes (carbohydrates and protein), fatty acids yield the most ATP on an energy per gram basis, when they are completely oxidized to CO_2 and water by β-oxidation and the citric acid cycle. Fatty acids (mainly in the form of triglycerides) are therefore the foremost storage form of fuel in most animals, and to a lesser extent in plants. In addition, fatty acids are important components of the phospholipids that form the phospholipid bilayers out of which all the membranes of the cell are constructed (the cell wall, and the membranes that enclose all the organelles within the cells, such as the nucleus, the mitochondria, endoplasmic reticulum, and the Golgi apparatus). Fatty acids can also be cleaved, or partially cleaved, from their chemical attachments in the cell membrane to form second messengers within the cell, and local hormones in the immediate vicinity of the cell. The prostaglandins made from arachidonic acid stored in the cell membrane, are probably the most well known group of these local hormones.

Fatty Acid Catabolism

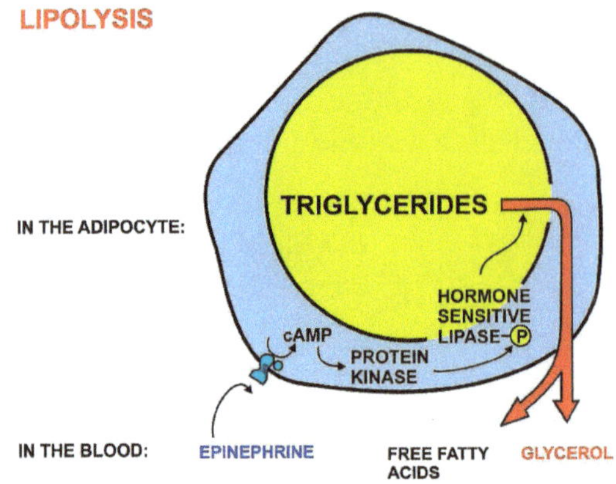

A diagrammatic illustration of the process of lipolysis (in a fat cell) induced by high epinephrine and low insulin levels in the blood. Epinephrine binds to a beta-adrenergic receptor in the cell wall of the adipocyte, which causes cAMP to be generated inside the cell. The cAMP activates a protein kinase, which phosphorylates and thus, in turn, activates a hormone-sensitive lipase in the fat cell. This lipase cleaves free fatty acids from their attachment to glycerol in the fat stored in the fat droplet of the adipocyte. The free fatty acids and glycerol are then released into the blood. However more recent studies have shown that adipose triglyceride lipase has to first convert triacylglycerides to diacylglycerides, and that hormone-sensitive lipase converts the diacylglycerides to monoglycerides and free fatty acids. Monoglycerides are hydrolyzed by monoglyceride lipase. The activity of hormone sensitive lipase is regulated by the circulation hormones insulin, glucagon, norepinephrine, and epinephrine, as shown in the diagram.

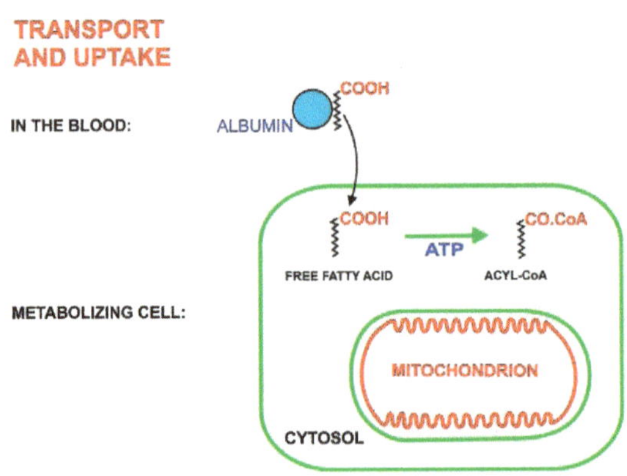

A diagrammatic illustration of the transport of free fatty acids in the blood attached to plasma albumin, its diffusion across the cell membrane using a protein transporter, and its activation, using ATP, to form acyl-CoA in the cytosol. The illustration is, for diagrammatic purposes, of a 12 carbon fatty acid. Most fatty acids in human plasma are

16 or 18 carbon atoms long.

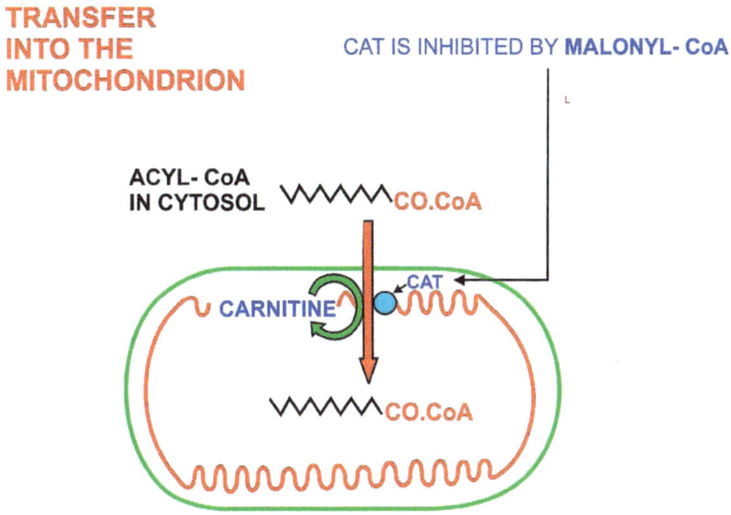

A diagrammatic illustration of the transfer of an acyl-CoA molecule across the inner membrane of the mitochondrion by carnitine-acyl-CoA transferase (CAT). The illustrated acyl chain is, for diagrammatic purposes, only 12 carbon atoms long. Most fatty acids in human plasma are 16 or 18 carbon atoms long. CAT is inhibited by high concentrations of malonyl-CoA (the first committed step in fatty acid synthesis) in the cytoplasm. This means that fatty acid synthesis and fatty acid catabolism cannot occur simultaneously in any given cell.

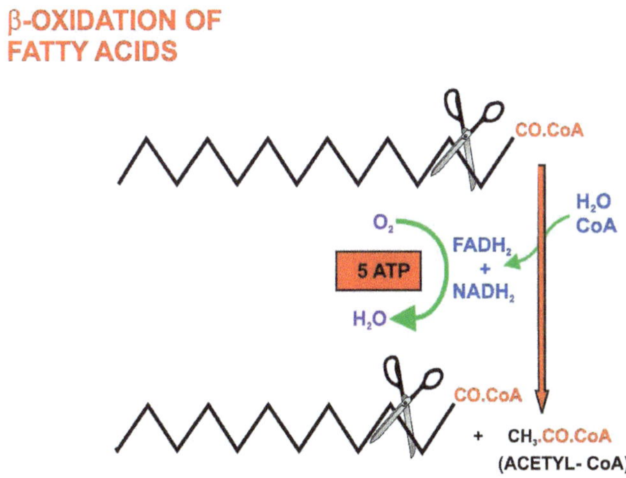

A diagrammatic illustration of the process of the beta-oxidation of an acyl-CoA molecule in the mitochodrial matrix. During this process an acyl-CoA molecule which is 2 carbons shorter than it was at the beginning of the process is formed. Acetyl-CoA, water and 5 ATP molecules are the other products of each beta-oxidative event, until the entire acyl-CoA molecule has been reduced to a set of acetyl-CoA molecules.

Fatty acids are released, between meals, from the fat depots in adipose tissue, where they are stored as triglycerides, as follows:

- Lipolysis, the removal of the fatty acid chains from the glycerol to which they are bound in their storage form as triglycerides (or fats), is carried out by lipases. These lipases are activated by high epinephrine and glucagon levels in the blood (or norepinephrine secreted by sympathetic nerves in adipose tissue), caused by declining blood glucose levels after meals, which simultaneously lowers the insulin level in the blood.
- Once freed from glycerol, the free fatty acids enter the blood, which transports them, attached to plasma albumin, throughout the body.
- Long chain free fatty acids enter the metabolizing cells (i.e. most living cells in the body except red blood cells and neurons in the central nervous system) through specific transport proteins, such as the SLC27 family fatty acid transport protein. Red blood cells do not contain mitochondria and are therefore incapable of metabolizing fatty acids; the tissues of the central nervous system cannot use fatty acids, despite containing mitochondria, because fatty acids cannot cross the blood brain barrier into the interstitial fluids that bathe these cells.
- Once inside the cell long-chain-fatty-acid—CoA ligase catalyzes the reaction between a fatty acid molecule with ATP (which is broken down to AMP and inorganic pyrophosphate) to give a fatty acyl-adenylate, which then reacts with free coenzyme A to give a fatty acyl-CoA molecule.
- In order for the acyl-CoA to enter the mitochondrion the carnitine shuttle is used:
- Acyl-CoA is transferred to the hydroxyl group of carnitine by carnitine palmitoyltransferase I, located on the cytosolic faces of the outer and inner mitochondrial membranes.
- Acyl-carnitine is shuttled inside by a carnitine-acylcarnitine translocase, as a carnitine is shuttled outside.
- Acyl-carnitine is converted back to acyl-CoA by carnitine palmitoyltransferase II, located on the interior face of the inner mitochondrial membrane. The liberated carnitine is shuttled back to the cytosol, as an acyl-CoA is shuttled into the matrix.
- Beta oxidation, in the mitochondrial matrix, then cuts the long carbon chains of the fatty acids (in the form of acyl-CoA molecules) into a series of two-carbon (acetate) units, which, combined with co-enzyme A, form molecules of acetyl CoA, which condense with oxaloacetate to form citrate at the "beginning" of the citric acid cycle. It is convenient to think of this reaction as marking the "starting point" of the cycle, as this is when fuel - acetyl-CoA - is added to the cycle, which will be dissipated as CO_2 and H_2O with the release of a substantial quantity of energy captured in the form of ATP, during the course of each turn of the cycle.

Briefly, the steps in β-oxidation (the initial breakdown of free fatty acids into acetyl-CoA) are as follows:

- Dehydrogenation by acyl-CoA dehydrogenase, yielding 1 $FADH_2$
- Hydration by enoyl-CoA hydratase
- Dehydrogenation by 3-hydroxyacyl-CoA dehydrogenase, yielding 1 NADH + H^+
- Cleavage by thiolase, yielding 1 acetyl-CoA and a fatty acid that has now been shortened by 2 carbons (forming a new, shortened acyl-CoA)

This β-oxidation reaction is repeated until the fatty acid has been completely reduced to acetyl-CoA or, in, the case of fatty acids with odd numbers of carbon atoms, acetyl-CoA and 1 molecule of propionyl-CoA per molecule of fatty acid. Each β-oxidative cut of the acyl-CoA molecule yields 5 ATP molecules.

- The acetyl-CoA produced by β-oxidation enters the citric acid cycle in the mitochondrion by combining with oxaloacetate to form citrate. This results in the complete combustion of the acetyl-CoA to CO_2 and water. The energy released in this process is captured in the form of 1 GTP and 11 ATP molecules per acetyl-CoA molecule oxidized. This is the fate of acetyl-CoA wherever β-oxidation of fatty acids occurs, except under certain circumstances in the liver. In the liver oxaloacetate can be wholly or partially diverted into the gluconeogenic pathway during fasting, starvation, a low carbohydrate diet, prolonged strenuous exercise, and in uncontrolled type 1 diabetes mellitus. Under these circumstances oxaloacetate is hydrogenated to malate which is then removed from the mitochondrion to be converted into glucose in the cytoplasm of the liver cells, from where it is released into the blood. In the liver, therefore, oxaloacetate is unavailable for condensation with acetyl-CoA when significant gluconeogenesis has been stimulated by low (or absent) insulin and high glucagon concentrations in the blood. Under these circumstances acetyl-CoA is diverted to the formation of acetoacetate and beta-hydroxybutyrate. Acetoacetate, beta-hydroxybutyrate, and their spontaneous breakdown product, acetone, are frequently, but confusingly, known as ketone bodies (as they are not "bodies" at all, but water-soluble chemical substances). The ketone bodies are released by the liver into the blood. All cells with mitochondria can take ketone bodies up from the blood and reconvert them into acetyl-CoA, which can then be used as fuel in their citric acid cycles, as no other tissue can divert its oxaloacetate into the gluconeogenic pathway in the way that this can occur in the liver. Unlike free fatty acids, ketone bodies can cross the blood-brain barrier and are therefore available as fuel for the cells of the central nervous system, acting as a substitute for glucose, on which these cells normally survive. The occurrence of high levels of ketone bodies in the blood during starvation, a low carbohydrate diet, prolonged heavy exercise and uncontrolled type 1 diabetes mellitus is known as ketosis, and, in its extreme form, in out-of-control type 1 diabetes mellitus, as ketoacidosis.

The glycerol released by lipase action is phosphorylated by glycerol kinase in the liver (the only tissue in which this reaction can occur), and the resulting glycerol 3-phosphate is oxidized to dihydroxyacetone phosphate. The glycolytic enzyme triose phosphate isomerase converts this compound to glyceraldehyde 3-phosphate, which is oxidized via glycolysis, or converted to glucose via gluconeogenesis.

Fatty Acids as an Energy Source

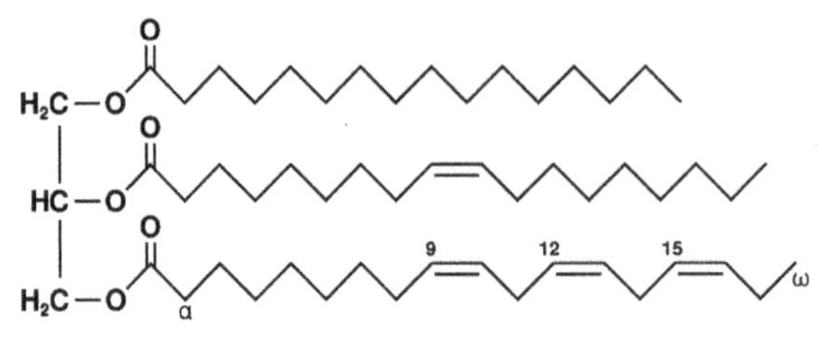

Example of an unsaturated fat triglyceride. Left part: glycerol, right part from top to bottom: palmitic acid, oleic acid, alpha-linolenic acid. Chemical formula: $C_{55}H_{98}O_6$

Fatty acids, stored as triglycerides in an organism, are an important source of energy because they are both reduced and anhydrous. The energy yield from a gram of fatty acids is approximately 9 kcal (37 kJ), compared to 4 kcal (17 kJ) for carbohydrates. Since the hydrocarbon portion of fatty acids is hydrophobic, these molecules can be stored in a relatively anhydrous (water-free) environment. Carbohydrates, on the other hand, are more highly hydrated. For example, 1 g of glycogen can bind approximately 2 g of water, which translates to 1.33 kcal/g (4 kcal/3 g). This means that fatty acids can hold more than six times the amount of energy per unit of storage mass. Put another way, if the human body relied on carbohydrates to store energy, then a person would need to carry 31 kg (67.5 lb) of hydrated glycogen to have the energy equivalent to 4.6 kg (10 lb) of fat.

Hibernating animals provide a good example for utilizing fat reserves as fuel. For example, bears hibernate for about 7 months, and, during this entire period, the energy is derived from degradation of fat stores. Migrating birds similarly build up large fat reserves before embarking on their intercontinental journeys. Thus the young adult human's fat stores average between about 10–20 kg, but varies greatly depending on age, gender, and individual disposition. By contrast the human body stores only about 400 g of glycogen, of which 300 g is locked inside the skeletal muscles and is unavailable to the body as a whole. The 100 g or so of glycogen stored in the liver is depleted within one day of starvation. Thereafter the glucose that is released into the blood by the liver for general use by the body tissues, has to be synthesized from the glucogenic amino acids and a few other gluconeogenic substrates, which do not include fatty acids.

Animals and Plants Synthesize Carbohydrates from both Glycerol and Fatty Acids

Fatty acids are broken down to acetyl-CoA by means of beta oxidation inside the mitochondria, whereas fatty acids are synthesized from acetyl-CoA outside the mitochondria, in the cytosol. The two pathways are distinct, not only in where they occur, but also in the reactions that occur, and the substrates that are used. The two pathways are mutually inhibitory, preventing the acetyl-CoA produced by beta-oxidation from entering the synthetic pathway via the acetyl-CoA carboxylase reaction. It can also not be converted to pyruvate as the pyruvate decarboxylase reaction is irreversible. Instead the acetyl-CoA produced by the beta-oxidation of fatty acids condenses with oxaloacetate, to enter the citric acid cycle. During each turn of the cycle, two carbon atoms leave the cycle as CO_2 in the decarboxylation reactions catalyzed by isocitrate dehydrogenase and alpha-ketoglutarate dehydrogenase. Thus each turn of the citric acid cycle oxidizes an acetyl-CoA unit while regenerating the oxaloacetate molecule with which the acetyl-CoA had originally combined to form citric acid. The decarboxylation reactions occur before malate is formed in the cycle. Only plants possess the enzymes to convert acetyl-CoA into oxaloacetate from which malate can be formed to ultimately be converted to glucose.

However acetyl-CoA can be converted to acetoacetate, which can decarboxylate to acetone (either spontaneously, or by acetoacetate decarboxylase). It can then be further metabolized to isopropanol which is excreted in breath/urine, or by CYP2E1 into hydroxyacetone (acetol). Acetol can be converted to propylene glycol. This converts to formate and acetate (the latter converting to glucose), or pyruvate (by two alternative enzymes), or propionaldehyde, or to L-lactaldehyde then L-lactate (the common lactate isomer). Another pathway turns acetol to methylglyoxal, then to pyruvate, or to D-lactaldehyde (via S-D-lactoyl-glutathione or otherwise) then D-lactate. D-lactate metabolism (to glucose) is slow or impaired in humans, so most of the D-lactate is excreted in the urine; thus D-lactate derived from acetone can contribute significantly to the metabolic acidosis associated with ketosis or isopropanol intoxication. L-Lactate can complete the net conversion of fatty acids into glucose. The first experiment to show conversion of acetone to glucose was carried out in 1951. This, and further experiments used carbon isotopic labelling. Up to 11% of the glucose can be derived from acetone during starvation in humans. The glycerol released into the blood during the lipolysis of triglycerides in adipose tissue can only be taken up by the liver. Here it is converted into glycerol 3-phosphate by the action of glycerol kinase which hydrolyzes one molecule of ATP per glycerol molecule which is phosphorylated. Glycerol 3-phosphate is then oxidized to dihydroxyacetone phosphate, which is, in turn, converted into glyceraldehyde 3-phosphate by the enzyme triose phosphate isomerase. From here the three carbon atoms of the original glycerol can be oxidized via glycolysis, or converted to glucose via gluconeogenesis.

Other Functions and uses of Fatty Acids

Intracellular Signaling

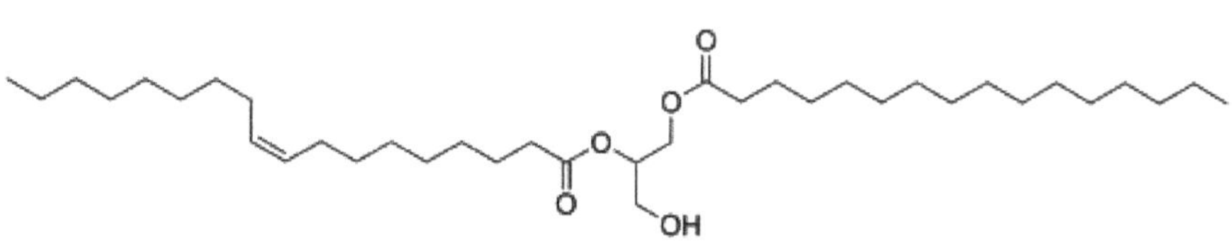

Chemical structure of the diglyceride 1-palmitoyl-2-oleoyl-glycerol.

Fatty acids are an integral part of the phospholipids that make up the bulk of the plasma membranes, or cell membranes, of cells. These phospholipids can be cleaved into diacylglycerol (DAG) and inositol trisphosphate (IP_3) through hydrolysis of the phospholipid, phosphatidylinositol 4,5-bisphosphate (PIP_2), by the cell membrane bound enzyme phospholipase C (PLC).

An example of a diacyl-glycerol shown on the right. This DAG is 1-palmitoyl-2-oleoyl-glycerol, which contains side-chains derived from palmitic acid and oleic acid. Diacylglycerols can also have many other combinations of fatty acids attached at either the C-1 and C-2 positions or the C-1 and C-3 positions of the glycerol molecule. 1,2 disubstituted glycerols are always chiral, 1,3 disubstituted glycerols are chiral if the substituents are different from each other.

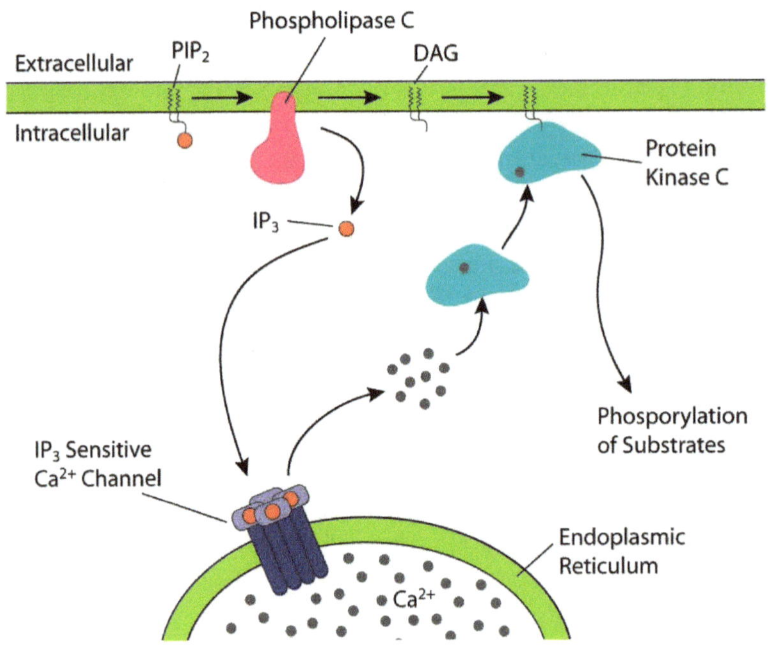

PIP_2 cleavage to IP_3 and DAG. IP_3initiates intracellular calcium release, while DAG activates PKC (protein kinase C). Note: PLC (phospholipase C) is not an intermediate, as possibly suggested by the diagram, but is the enzyme that catalyzes the IP3/DAG separation.

Inositol trisphosphate (IP_3) functions as an intracellular second messenger, which initiates the intracellular release of calcium ions (which activates intracellular enzymes, causes the release of hormones and neurotransmitters from the cells in which they are stored, and causes smooth muscle contraction when released by IP_3), and the activation of protein kinase C (PKC), which is then translocated from the cell cytoplasm to the cell membrane. Although inositol trisphosphate, (IP_3), diffuses into the cytosol, diacylglycerol (DAG) remains within the plasma membrane, due to its hydrophobic properties. IP_3 stimulates the release of calcium ions from the smooth endoplasmic reticulum, whereas DAG is a physiological activator of protein kinase C (PKC), promoting its translocation from the cytosol to the plasma membrane. PKC is a multifunctional protein kinase which phosphorylates serine and threonine residues in many target proteins. However PKC is only active in the presence of calcium ions, and it is DAG that increases the affinity of PKC for Ca^{2+} and thereby renders it active at the physiological intracellular levels of this ion.

Diacylglycerol and IP_3 act transiently because both are rapidly metabolized. This is important as their message function should not linger after the message has been" received" by their target molecules. DAG can be phosphorylated to phosphatidate or it can be it can be hydrolysed to glycerol and its constituent fatty acids. IP_3 is rapidly converted to into derivatives that that do not open calcium ion channels.

Eicosanoid Paracrine Hormones

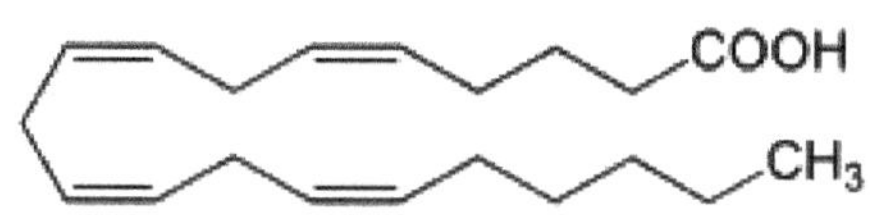

Arachidonic acid.

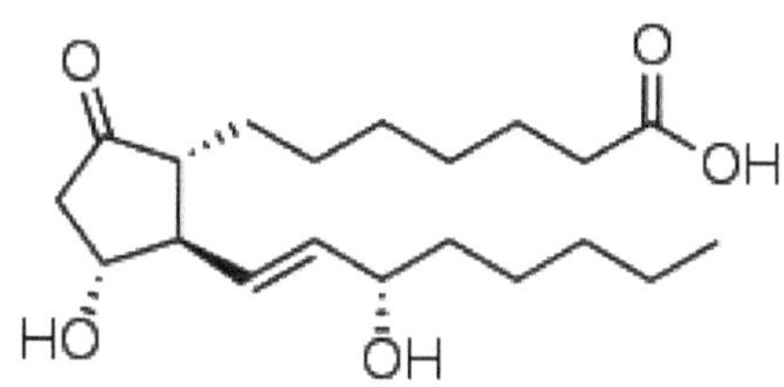

Prostaglandin E_1 - Alprostadil.

The prostaglandins are a group of physiologically active lipid compounds having diverse hormone-like effects in animals. Prostaglandins have been found in almost every tissue in humans and other animals. They are enzymatically derived from arachidonic acid a 20-carbon polyunsaturated fatty acid. Every prostaglandin therefore contains 20 carbon atoms, including a 5-carbon ring. They are a subclass of eicosanoids and form the prostanoid class of fatty acid derivatives.

The prostaglandins are synthesized in the cell membrane by the cleavage of arachidonate from the phospholipids that make up the membrane. This is catalyzed either by phospholipase A_2 acting directly on a membrane phospholipid, or by a lipase acting on DAG (diacyl-glycerol). The arachidonate is then acted upon by the cyclooxygenase component of prostaglandin synthase. This forms a cyclopentane ring in roughly the middle of the fatty acid chain. The reaction also adds 4 oxygen atoms derived from two molecules of O_2. The resulting molecule is prostaglandin G_2 which is converted by the hydroperoxidase component of the enzyme complex into prostaglandin H_2. This highly unstable compound is rapidly transformed into other prostaglandins, prostacyclin and thromboxanes. These are then released into the interstitial fluids surrounding the cells that have manufactured the eicosanoid hormone.

If arachidonate is acted upon by a lipoxygenase instead of cyclooxygenase, Hydroxyeicosatetraenoic acids and leukotrienes are formed. They also act as local hormones.

Prostaglandins were originally believed to leave the cells via passive diffusion because of their high lipophilicity. The discovery of the prostaglandin transporter (PGT, SLCO2A1), which mediates the cellular uptake of prostaglandin, demonstrated that diffusion alone cannot explain the penetration of prostaglandin through the cellular membrane. The release of prostaglandin has now also been shown to be mediated by a specific transporter, namely the multidrug resistance protein 4 (MRP4, ABCC4), a member of the ATP-binding cassette transporter superfamily. Whether MRP4 is the only transporter releasing prostaglandins from the cells is still unclear.

The structural differences between prostaglandins account for their different biological activities. A given prostaglandin may have different and even opposite effects in different tissues. The ability of the same prostaglandin to stimulate a reaction in one tissue and inhibit the same reaction in another tissue is determined by the type of receptor to which the prostaglandin binds. They act as autocrine or paracrine factors with their target cells present in the immediate vicinity of the site of their secretion. Prostaglandins differ from endocrine hormones in that they are not produced at a specific site but in many places throughout the human body. Prostaglandins have two derivatives: prostacyclins and thromboxanes. Prostacyclins are powerful locally acting vasodilators and inhibit the aggregation of blood platelets. Through their role in vasodilation, prostacyclins are also involved in inflammation. They are synthesized in the walls of blood vessels and serve the physiological function of preventing needless clot formation, as well as regulating the contraction of smooth muscle tissue. Conversely, thromboxanes (produced by platelet cells) are vasoconstrictors and facilitate platelet aggregation. Their name comes from their role in clot formation (thrombosis).

Dietary sources of Fatty Acids, their Digestion, Absorption, Transport in the Blood and Storage

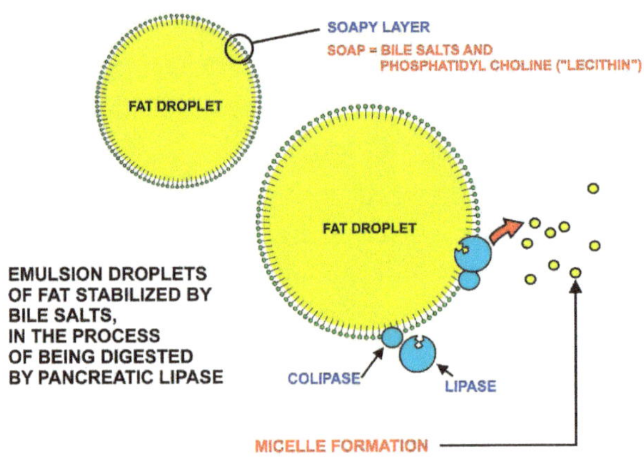

Dietary fats are emulsified in the duodenum by soaps in the form of bile salts and phospholipids, such as phosphatidylcholine. The fat droplets thus formed can be attacked by pancreatic lipase.

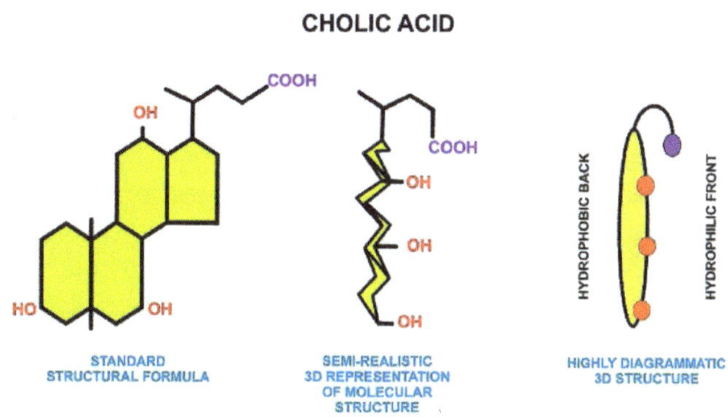

Structure of a bile acid (cholic acid), represented in the standard form, a semi-realistic 3D form, and a diagrammatic 3D form

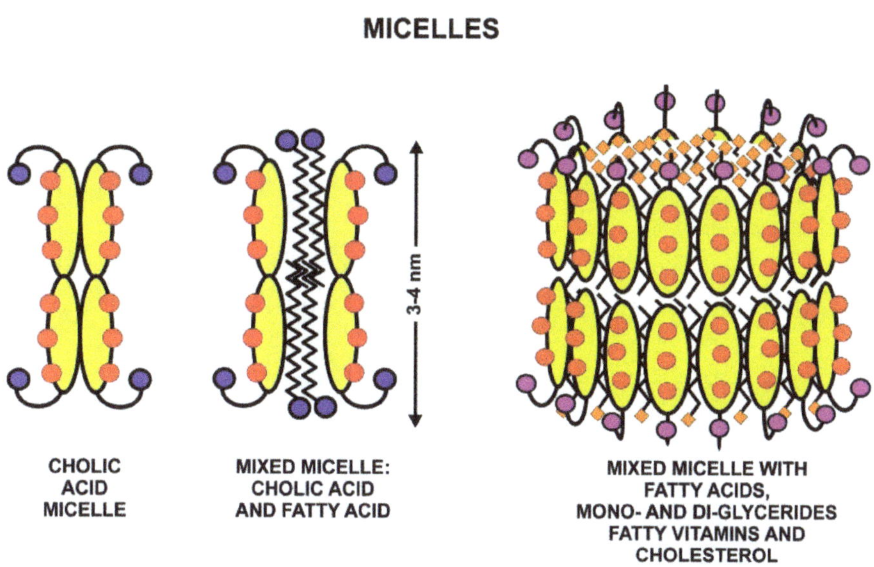

Diagrammatic illustration of mixed micelles formed in the duodenum in the presence of bile acids (e.g. cholic acid) and the digestion products of fats, the fat soluble vitamins and cholesterol.

A significant proportion of the fatty acids in the body are obtained from the diet, in the form of triglycerides of either animal or plant origin. The fatty acids in the fats obtained from land animals tend to be saturated, whereas the fatty acids in the triglycerides of fish and plants are often polyunsaturated and therefore present as oils.

These triglycerides, cannot be absorbed by the intestine. They are broken down into mono- and di-glycerides plus free fatty acids (but no free glycerol) by pancreatic lipase, which forms a 1:1 complex with a protein called colipase (also a constituent of pancreatic juice), which is necessary for its activity. The activated complex can work only at a water-fat interface. Therefore, it is essential that fats are first emulsified by bile salts for optimal activity of these enzymes. The digestion products consisting of a mixture of tri-, di- and monoglycerides and free fatty acids, which, together with the other fat soluble contents of the diet (e.g. the fat soluble vitamins and cholesterol) and bile salts form mixed micelles, in the watery duodenal contents (see diagrams on the left).

The contents of these micelles (but not the bile salts) enter the enterocytes (epithelial cells lining the small intestine) where they are resynthesized into triglycerides, and packaged into chylomicrons which are released into the lacteals (the capillaries of the lymph system of the intestines). These lacteals drain into the thoracic duct which empties into the venous blood at the junction of the left jugular and left subclavian veins on the lower left hand side of the neck. This means that the fat soluble products of digestion are discharged directly into the general circulation, without first passing through the liver, as all other digestion products do. The reason for this peculiarity is unknown.

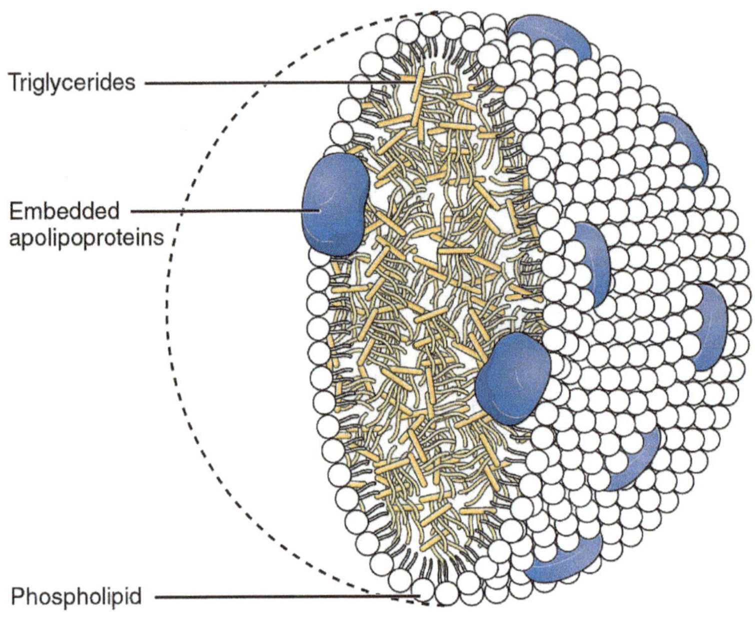

A shematic diagram of a chylomicron.

The chylomicrons circulate throughout the body, giving the blood plasma a milky, or creamy appearance after a fatty meal. Lipoprotein lipase on the endothelial surfaces of the capillaries, especially in adipose tissue, but to a lesser extent also in other tissues, partially digests the chylomicrons into free fatty acids, glycerol and chylomicron remnants. The fatty acids are absorbed by the adipocytes, but the glycerol and chylomicron remnants remain in the blood plasma, ultimately to be removed from the circulation by the liver.

The free fatty acids released by the digestion of the chylomicrons are absorbed by the adipocytes, where they are resynthesized into triglycerides using glycerol derived from glucose in the glycolytic pathway. These triglycerides are stored, until needed for the fuel requirements of other tissues, in the fat droplet of the adipocyte.

The liver absorbs a proportion of the glucose from the blood in the portal vein coming from the intestines. After the liver has replenished its glycogen stores (which amount to only about 100 g of glycogen when full) much of the rest of the glucose is converted into fatty acids as described below. These fatty acids are combined with glycerol to form triglycerides which are packaged into droplets very similar to chylomicrons, but known as very low-density lipoproteins (VLDL). These VLDL droplets are handled in exactly the same manner as chylomicrons, except that the VLDL remnant is known as an intermediate-density lipoprotein, (IDL), which is capable of scavenging cholesterol from the blood. This converts IDL into low-density lipoprotein, (LDL), which is taken up by cells that require cholesterol for incorporation into their cell membranes or for synthetic purposes (e.g. the formation of the steroid hormones). The remainder of the LDLs is removed by the liver.

Adipose tissue and lactating mammary glands also take up glucose from the blood for conversion into triglycerides. This occurs in the same way as it does in the liver, except that these tissues do not release the triglycerides thus produced as VLDL into the blood. Adipose tissue cells store the triglycerides in their fat droplets, ultimately to release them again as free fatty acids and glycerol into the blood (as described above), when the plasma concentration of insulin is low, and that of glucagon and/or epinephrine is high. Mammary glands discharge the fat (as cream fat droplets) into the milk that they produce under the influence of the anterior pituitary hormone prolactin.

All cells in the body need to manufacture and maintain their cell walls and the membranes of their organelles. Whether they rely for this entirely on free fatty acids absorbed from the blood, or are able to synthesize their own fatty acids from the blood glucose, is not known. The cells of the central nervous system will almost certainly have the capability of manufacturing their own fatty acids, as these molecules cannot reach them through the blood brain barrier, while, on the other hand, no cell in the body can manufacture the required essential fatty acids which have to be obtained from the diet and delivered to each cell via the blood.

Fatty Acid Synthesis

Much like β-oxidation, straight-chain fatty acid synthesis occurs via the six recurring reactions shown below, until the 16-carbon palmitic acid is produced.

The diagrams presented show how fatty acids are synthesized in microorganisms and list the enzymes found in Escherichia coli. These reactions are performed by fatty acid synthase II (FASII), which in general contain multiple enzymes that act as one complex. FASII is present in prokaryotes, plants, fungi, and parasites, as well as in mitochondria.

In animals, as well as some fungi such as yeast, these same reactions occur on fatty acid synthase I (FASI), a large dimeric protein that has all of the enzymatic activities required to create a fatty acid. FASI is less efficient than FASII; however, it allows for the formation of more molecules, including "medium-chain" fatty acids via early chain termination.

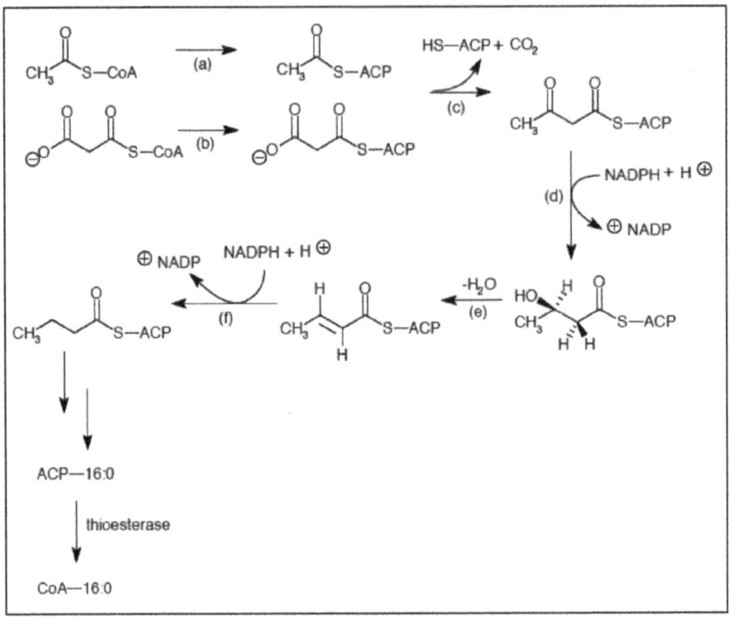

Synthesis of saturated fatty acids via Fatty Acid Synthase II in E. coli.

Once a 16:0 carbon fatty acid has been formed, it can undergo a number of modifications, resulting in desaturation and/or elongation. Elongation, starting with stearate (18:0), is performed mainly in the ER by several membrane-bound enzymes. The enzymatic steps involved in the elongation process are principally the same as those carried out by FAS, but the four principal successive steps of the elongation are performed by individual proteins, which may be physically associated.

Step	Enzyme	Reaction	Description
(a)	Acetyl CoA:ACP transacylase	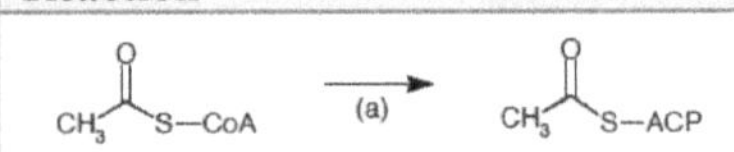	Activates acetyl CoA for reaction with malonyl-ACP
(b)	Malonyl CoA:ACP transacylase	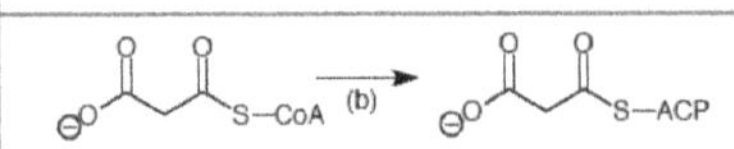	Activates malonyl CoA for reaction with acetyl-ACP
(c)	3-ketoacyl-ACP synthase	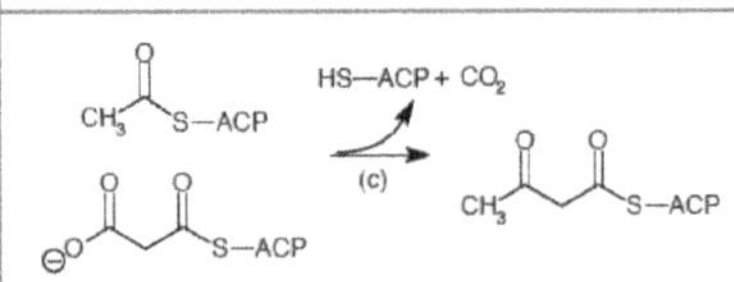	Reacts priming acetyl-ACP with chain-extending malonyl-ACP.
(d)	3-ketoacyl-ACP reductase	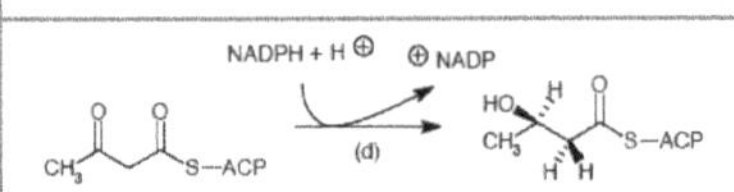	Reduces the carbon 3 ketone to a hydroxyl group
(e)	3-Hydroxyacyl ACP dehydrase	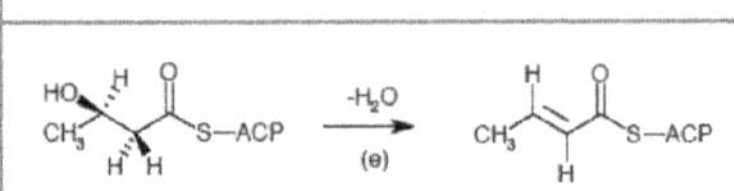	Removes water

(f)	Enoyl-ACP reductase	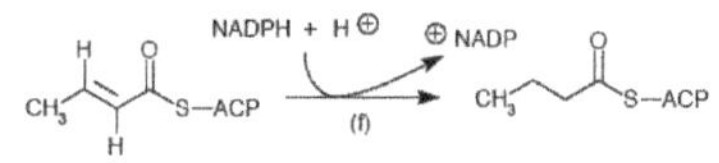	Reduces the C2-C3 double bond.

Abbreviations: ACP – Acyl carrier protein, CoA – Coenzyme A, NADP – Nicotinamide adenine dinucleotide phosphate.

Note that during fatty synthesis the reducing agent is NADPH, whereas NAD is the oxidizing agent in beta-oxidation (the breakdown of fatty acids to acetyl-CoA). This difference exemplifies a general principle that NADPH is consumed during biosynthetic reactions, whereas NADH is generated in energy-yielding reactions. (Thus NADPH is also required for the synthesis of cholesterol from acetyl-CoA; while NADH is generated during glycolysis.) The source of the NADPH is two-fold. When malate is oxidatively decarboxylated by "$NADP^+$-linked malic enzyme" pyruvate, CO_2 and NADPH are formed. NADPH is also formed by the pentose phosphate pathway which converts glucose into ribose, which can be used in synthesis of nucleotides and nucleic acids, or it can be catabolized to pyruvate.

Glycolytic end Products are used in the Conversion of Carbohydrates into Fatty Acids

In humans, fatty acids are formed from carbohydrates predominantly in the liver and adipose tissue, as well as in the mammary glands during lactation. The cells of the central nervous system probably also make most of the fatty acids needed for the phospholipids of their extensive membranes from glucose, as blood-born fatty acids cannot cross the blood brain barrier to reach these cells. However, how the essential fatty acids, which mammals cannot synthesize themselves, but are nevertheless important components of cell membranes (and other functions described above) reach them is unknown.

The pyruvate produced by glycolysis is an important intermediary in the conversion of carbohydrates into fatty acids and cholesterol. This occurs via the conversion of pyruvate into acetyl-CoA in the mitochondrion. However, this acetyl CoA needs to be transported into cytosol where the synthesis of fatty acids and cholesterol occurs. This cannot occur directly. To obtain cytosolic acetyl-CoA, citrate (produced by the condensation of acetyl CoA with oxaloacetate) is removed from the citric acid cycle and carried across the inner mitochondrial membrane into the cytosol. There it is cleaved by ATP citrate lyase into acetyl-CoA and oxaloacetate. The oxaloacetate is returned to mitochondrion as malate (and then converted back into oxaloacetate to transfer more acetyl-CoA out of the mitochondrion). The cytosolic acetyl-CoA is carboxylated by acetyl CoA carboxylase into malonyl CoA, the first committed step in the synthesis of fatty acids.

Regulation of Fatty Acid Synthesis

Acetyl-CoA is formed into malonyl-CoA by acetyl-CoA carboxylase, at which point malonyl-CoA is destined to feed into the fatty acid synthesis pathway. Acetyl-CoA carboxylase is the point of regulation in saturated straight-chain fatty acid synthesis, and is subject to both phosphorylation and allosteric regulation. Regulation by phosphorylation occurs mostly in mammals, while allosteric regulation occurs in most organisms. Allosteric control occurs as feedback inhibition by palmitoyl-CoA and activation by citrate. When there are high levels of palmitoyl-CoA, the final product of saturated fatty acid synthesis, it allosterically inactivates acetyl-CoA carboxylase to prevent a build-up of fatty acids in cells. Citrate acts to activate acetyl-CoA carboxylase under high levels, because high levels indicate that there is enough acetyl-CoA to feed into the Krebs cycle and produce energy.

High plasma levels of insulin in the blood plasma (e.g. after meals) cause the dephosphorylation of acetyl-CoA carboxylase, thus promoting the formation of malonyl-CoA from acetyl-CoA, and consequently the conversion of carbohydrates into fatty acids, while epinephrine and glucagon (released into the blood during starvation and exercise) cause the phosphorylation of this enzyme, inhibiting lipogenesis in favor of fatty acid oxidation via beta-oxidation.

Disorders

Disorders of fatty acid metabolism can be described in terms of, for example, hypertriglyceridemia (too high level of triglycerides), or other types of hyperlipidemia. These may be familial or acquired. Familial types of disorders of fatty acid metabolism are generally classified as inborn errors of lipid metabolism. These disorders may be described as fatty oxidation disorders or as a *lipid storage disorders*, and are any one of several inborn errors of metabolism that result from enzyme defects affecting the ability of the body to oxidize fatty acids in order to produce energy within muscles, liver, and other cell types.

ALDOL CONDENSATION

An **aldol condensation** is a condensation reaction in organic chemistry in which an enol or an enolate ion reacts with a carbonyl compound to form a β-hydroxyaldehyde or β-hydroxyketone, followed by dehydration to give a conjugatedenone.

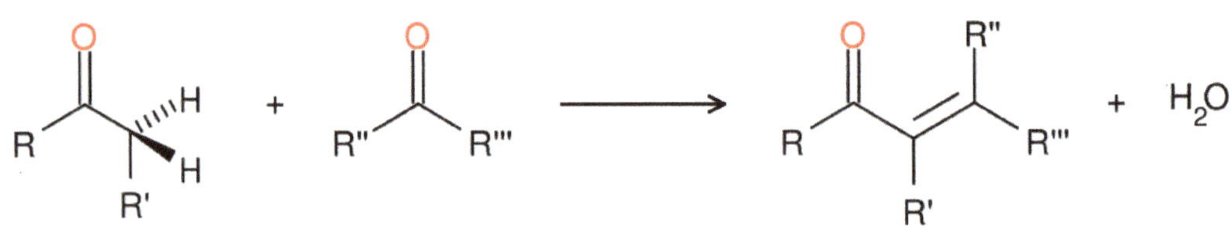

Aldol condensations are important in organic synthesis, because they provide a good way to form carbon–carbon bonds. For example, the Robinson annulation reaction sequence features an aldol condensation; the Wieland-Miescher ketone product is an important starting material for many organic syntheses. Aldol condensations are also commonly discussed in university level organic chemistry classes as a good bond-forming reaction that demonstrates important reaction mechanisms. In its usual form, it involves the nucleophilic addition of a ketone enolate to an aldehyde to form a β-hydroxy ketone, or "**aldol**" (**ald**ehyde + alcoh**ol**), a structural unit found in many naturally occurring molecules and pharmaceuticals.

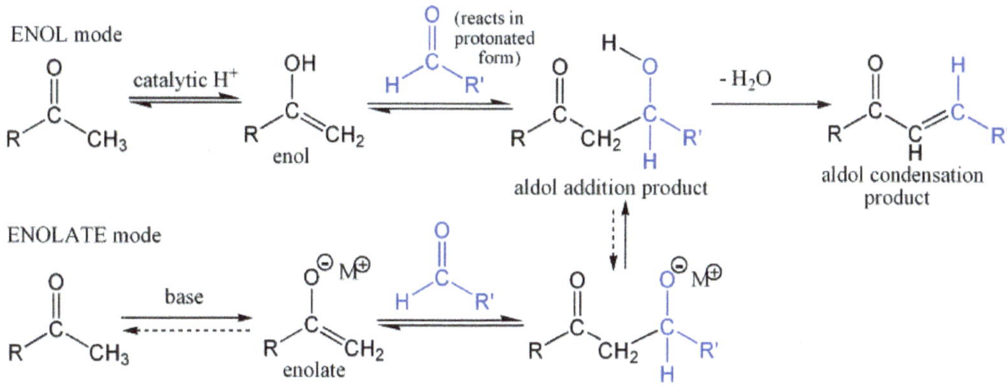

The name **aldol condensation** is also commonly used, especially in biochemistry, to refer to just the first (addition) stage of the process—the aldol reaction itself—as catalyzed by aldolases. However, the aldol reaction is not formally a condensation reaction because it does not involve the loss of a small molecule.

The reaction between an aldehyde/ketone and an aromatic carbonyl compound lacking an alpha-hydrogen (cross aldol condensation) is called the **Claisen-Schmidt condensation**. This reaction is named after two of its pioneering investigators Rainer Ludwig Claisen and J. G. Schmidt, who independently published on this topic in 1880 and 1881. An example is the synthesis of dibenzylideneacetone. Quantitative yields in Claisen-Schmidt reactions have been reported in the absence of solvent using sodium hydroxide as the base and plus benzaldehydes.

Mechanism

The first part of this reaction is an aldol reaction, the second part a dehydration—an elimination reaction (Involves removal of a water molecule or an alcohol molecule). Dehydration may be accompanied by decarboxylation when an activated carboxyl group is present. The aldol addition product can be dehydrated via two mechanisms; a strong base like potassium *t*-butoxide, potassium hydroxide or sodium hydride in an enolate mechanism, or in an acid-catalyzed enol mechanism. Depending on the nature of the desired product, the aldol condensation may be carried out under two broad types of conditions: kinetic control or thermodynamic control.

Base catalyzed aldol reaction (shown using $^{-}OCH_3$ as base)

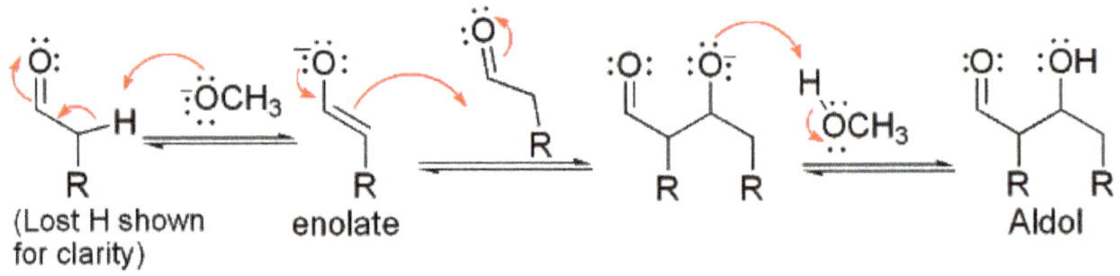

Base catalyzed dehydration (sometimes written as a single step)

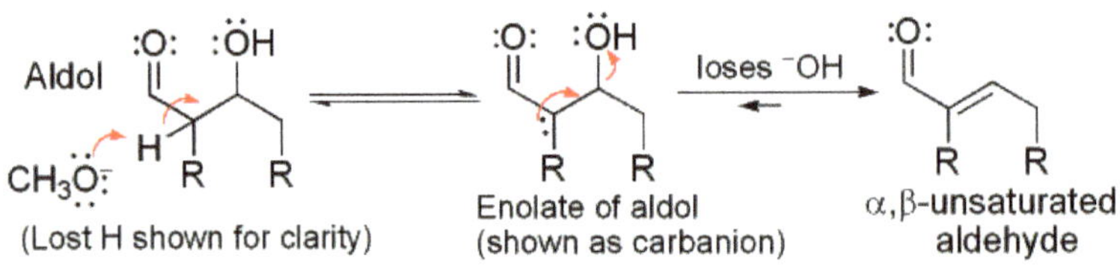

Acid catalyzed aldol reaction

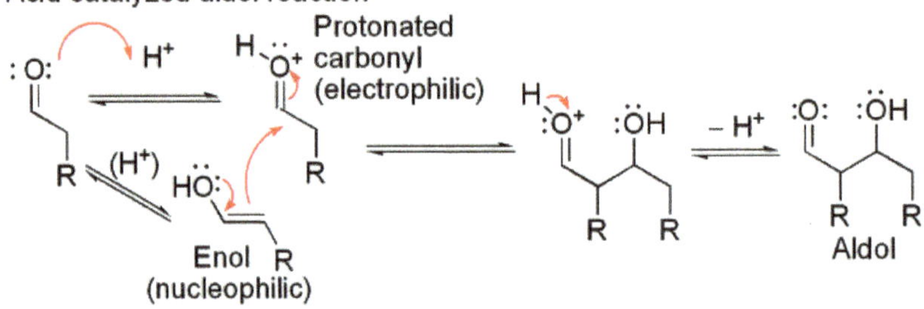

Acid catalyzed dehydration

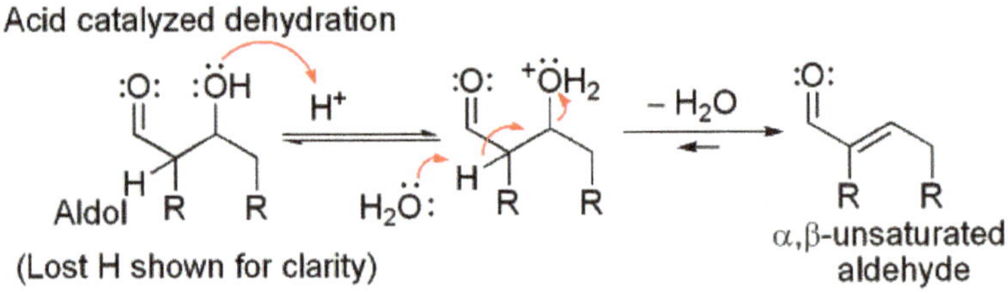

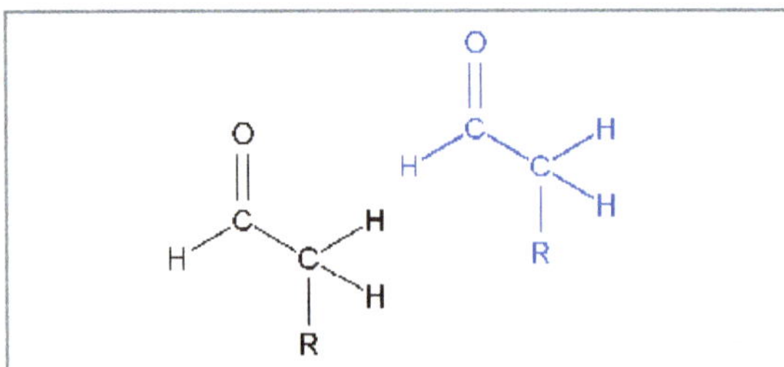	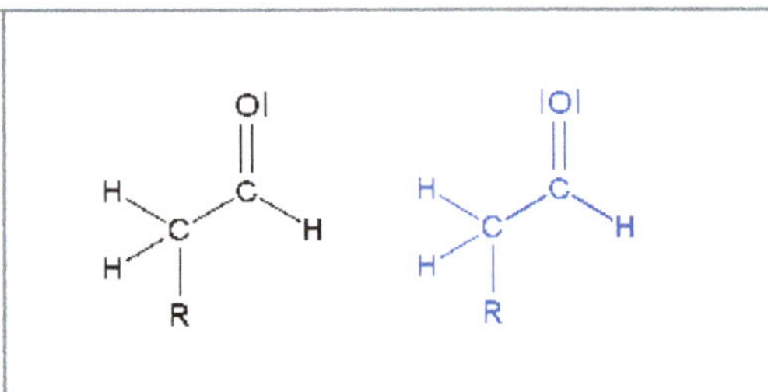
animation, base catalyzed	animation, acid catalyzed

Condensation Types

It is important to distinguish the aldol condensation from other addition reactions of carbonyl compounds.

- When the base is an amine and the active hydrogen compound is sufficiently activated the reaction is called a Knoevenagel condensation.
- In a Perkin reaction the aldehyde is aromatic and the enolate generated from an anhydride.
- A Claisen condensation involves two ester compounds.
- A Dieckmann condensation involves two ester groups in the *same molecule* and yields a cyclic molecule
- A Henry reaction involves an aldehyde and an aliphatic nitro compound.
- A Robinson annulation involves a α,β-unsaturated ketone and a carbonyl group, which first engage in a Michael reaction prior to the aldol condensation.
- In the Guerbet reaction, an aldehyde, formed *in situ* from an alcohol, self-condenses to the dimerized alcohol.
- In the Japp–Maitland condensation water is removed not by an elimination reaction but by a nucleophilic displacement

Aldox Process

In industry the **Aldox process** developed by Royal Dutch Shell and Exxon, converts propylene and syngas directly to 2-ethylhexanol via hydroformylation to butyraldehyde, aldol condensation to 2-ethylhexenal and finally hydrogenation.

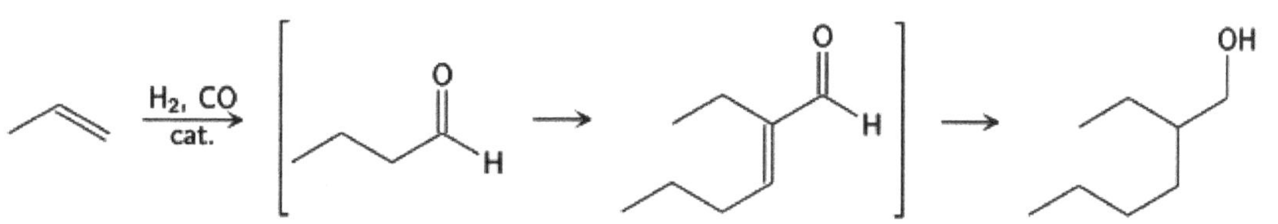

In one study crotonaldehyde is directly converted to 2-ethylhexanal in a palladium / Amberlyst / supercritical carbon dioxide system:

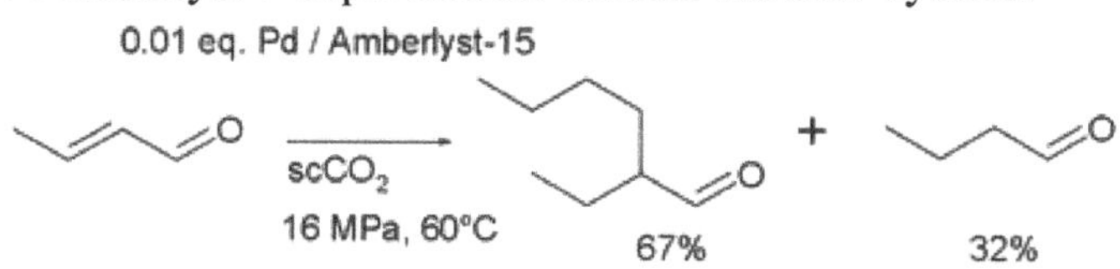

Scope

Ethyl 2-methylacetoacetate and campholenic aldehyde react in an Aldol condensation. The synthetic procedure is typical for this type of reaction. In the process, in addition to water, an equivalent of ethanol and carbon dioxide are lost in decarboxylation.

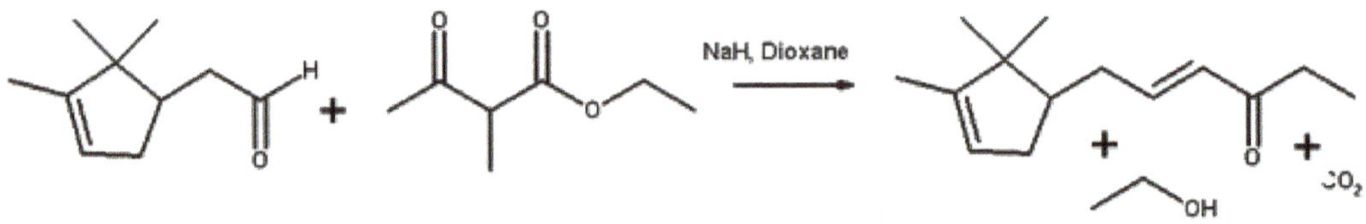

Ethyl glyoxylate **2** and diethyl 2-methylglutaconate **1** react to *isoprenetricarboxylic acid* **3** (isoprene skeleton) with sodium ethoxide. This reaction product is very unstable with initial loss of carbon dioxide and followed by many secondary reactions. This is believed to be due to steric strain resulting from the methyl group and the carboxylic group in the *cis*-dienoid structure.

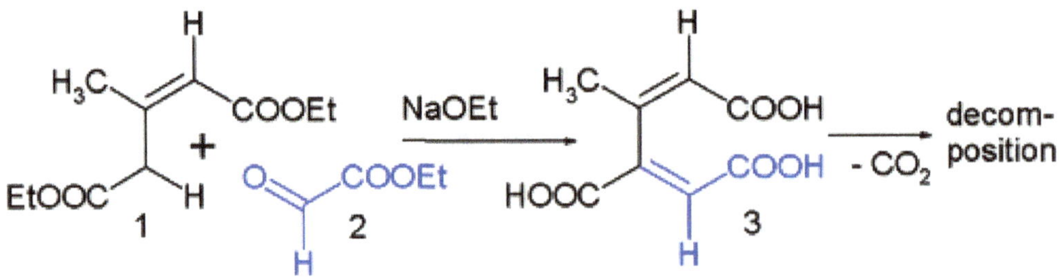

Occasionally an aldol condensation is buried in a multistep reaction or in catalytic cycle such as the one sketched below:

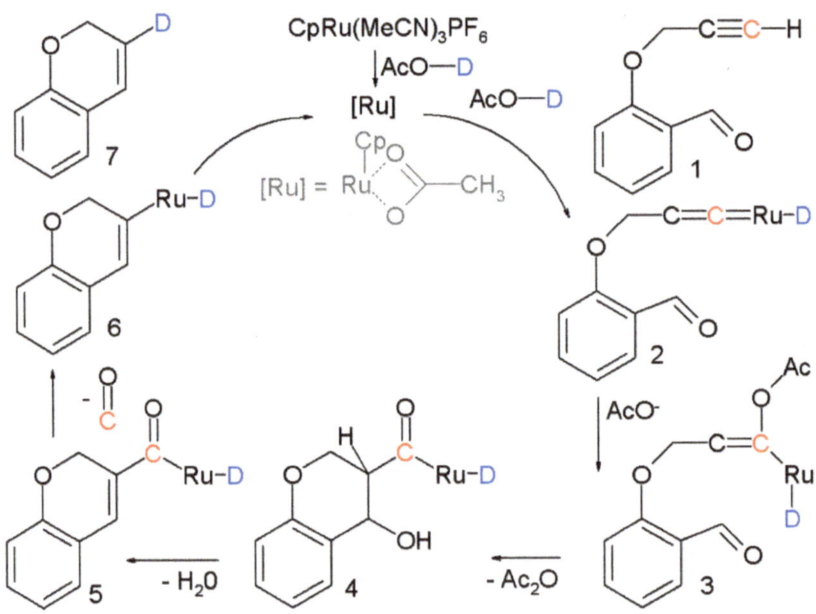

In this reaction an *alkynal* **1** is converted into a cycloalkene **7** with a ruthenium catalyst and the actual condensation takes place with intermediate **3** through **5**. Support for the reaction mechanism is based on isotope labeling.

The reaction between menthone and anisaldehyde is complicated due to steric shielding of the ketone group. This obstacle is overcome by using a strong base such as potassium hydroxide and a very polar solvent such as DMSO in the reaction below:

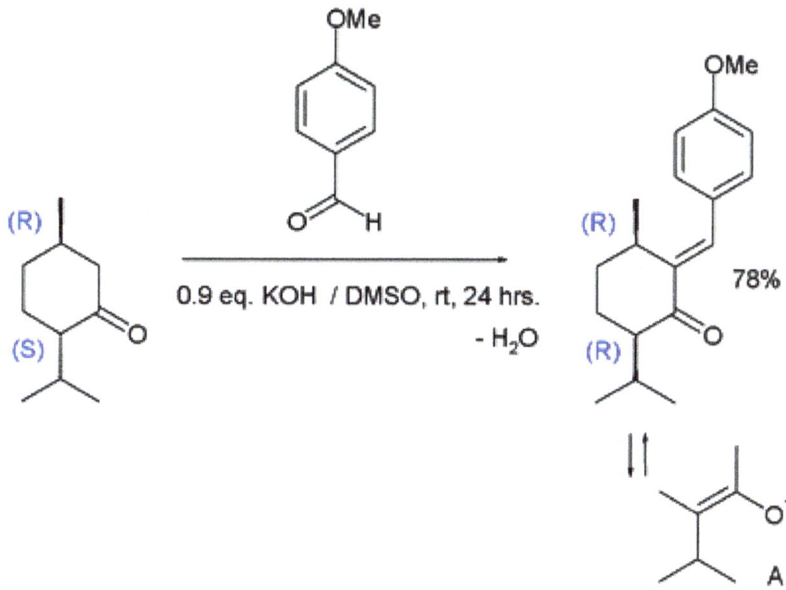

Due to epimerization through a common enolate ion (intermediate **A**) the reaction product has (*R*,*R*)-*cis*-configuration and not (*R*,*S*)-*trans*-configuration as in the starting material. Because it is only the *cis* isomer that precipitates from solution, this product is formed exclusively.

REFERENCES

1. Carey, Francis A.; Sundberg, Richard J. (1993). Advanced Organic Chemistry Part B Reactions and Synthesis (3rd ed.). 233 Spring Street ,NY: Plenum. p. 55. ISBN 0-306-43440-7.
2. Dzierba, C. D.; Zandi, K. S.; Moellers, T.; Shea, K. J. An Ascending Synthesis of Adrenalcorticosteroids. The Total Synthesis of (+)-Adrenosterone. J. Am. Chem. Soc. **1996**, 118, 4711-4712.
3. Wade, L. G. (2005). Organic Chemistry (6th ed.). Upper Saddle River, NJ: Prentice Hall. pp. 1056–1066. ISBN 0-13-236731-9.
4. Smith, M. B.; March, J. (2001). Advanced Organic Chemistry (5th ed.). New York: Wiley Interscience. pp. 1218–1223. ISBN 0-471-58589-0.
5. Mahrwald, R. (2004). Modern Aldol Reactions. **1, 2**. Weinheim, Germany: Wiley-VCH. pp. 1218–1223. ISBN 3-527-30714-1.
6. Heathcock, C. H. (1991). Additions to C-X π-Bonds, Part 2. Comprehensive Organic Synthesis. Selectivity, Strategy and Efficiency in Modern Organic Chemistry. **2**. Oxford: Pergamon. pp. 133–179. ISBN 0-08-040593-2.

7. Mukaiyama T. (1982). "The Directed Aldol Reaction". Organic Reactions. **28**: 203–331. doi:10.1002/0471264180.or028.03.

8. Paterson, I. (1988). "New Asymmetric Aldol Methodology Using Boron Enolates". Chemistry and Industry. London: Paterson Group. **12**: 390–394.

9. Claisen, L.; Claparède, A. (1881). "Condensationen von Ketonen mit Aldehyden". Berichte der Deutschen Chemischen Gesellschaft. **14** *(1): 2460–2468. doi:10.1002/cber.188101402192.*

10. Schmidt, J. G. (1881). "Ueber die Einwirkung von Aceton auf Furfurol und auf Bittermandelöl in Gegenwart von Alkalilauge". Berichte der Deutschen Chemischen Gesellschaft. **14** *(1): 1459–1461. doi:10.1002/cber.188101401306.*

11. March, J. (1985). Advanced Organic Chemistry: Reactions, Mechanisms and Structure (3rd ed.). Wiley Interscience. ISBN 0-471-85472-7.

12. Molecules 2012, 17, 571-583; doi:10.3390/molecules17010571

13. Nielsen, A. T.; Houlihan., W. J. (1968). "The Aldol Condensation". Organic Reactions. **16**: 1–438. doi:10.1002/0471264180.or016.01.

14. Carey, Francis A.; Sundberg, Richard J. (1993). Advanced Organic Chemistry Part A: Structure and Mechanisms (3rd ed.). 233 Spring Street, New York, N.Y.: Plenum. p. 458. ISBN 0-306-43440-7.

15. For example BG 881979

16. Seki, T.; Grunwaldt, J.-D.; Baiker, A. (2007). "Continuous catalytic "one-pot" multi-step synthesis of 2-ethylhexanal from crotonaldehyde". Chemical Communications. **2007** *(34): 3562–3564.* doi:10.1039/b710129e.

17. Badía, C.; Castro, J. M.; Linares-Palomino, P. J.; Salido, S.; Altarejos, J.; Nogueras, M.; Sánchez, A. (2004). "(E)-6-(2,2,3-Trimethyl-cyclopent-3-enyl)-hex-4-en-3-one" (pdf). Molbank. **2004** *(1): M388.* doi:10.3390/M388.

18. Ethyl 2-methylacetoacetate (**2**) is added to a stirred solution of sodium hydride in dioxane. Then campholenic aldehyde (**1**) is added and the mixture refluxed for 15 h. Then 2N hydrochloric acid is added and the mixture extracted with diethyl ether. The combined organic layers are washed with 2N hydrochloric acid, saturated sodium bicarbonate and brine. The organic phase is dried over anhydrous sodium sulfate and the solvent evaporated under reduced pressure to yield a residue that was purified by vacuum distillation to give **3** (58%).

19. Goren, M. B.; Sokoloski, E. A.; Fales, H. M. (2005). "2-Methyl-(1Z,3E)-butadiene-1,3,4-tricarboxylic Acid, "Isoprenetricarboxylic Acid"". Journal of Organic Chemistry. **70** *(18): 7429–7431.* doi:10.1021/jo0507892. PMID 16122270.

20. Varela, J. A.; Gonzalez-Rodriguez, C.; Rubin, S. G.; Castedo, L.; Saa, C. (2006). Ru-Catalyzed Cyclization of Terminal Alkynals to Cycloalkenes. Journal of the American Chemical Society. **128** *(30): 9576–9577.* doi:10.1021/ja0610434. PMID 16866480.

21. The ruthenium catalyst, $[CpRu(CH_3CN)_3]PF_6$, has a cyclopentadienyl ligand, three acetonitrile ligands and a phosphorus hexafluoride counterion; the acidic proton in the solvent (acetic acid) is replaced by deuterium for isotopic labeling.

Reaction conditions: 90°C, 24 hrs. 80% chemical yield. The first step is formation of the Transition metal carbene complex **2**. Acetic acid adds to this intermediate in a nucleophilic addition to form enolate **3** followed by aldol condensation to **5** at which stage a molecule of carbon monoxide is lost to **6**. The final step is reductive elimination to form the cycloalkene.

22. Vashchenko, V.; Kutulya, L.; Krivoshey, A. (2007). "Simple and Effective Protocol for Claisen-Schmidt Condensation of Hindered Cyclic Ketones with Aromatic Aldehydes". Synthesis. **2007** *(14): 2125–2134.* doi:10.1055/s-2007-983746.

CHAPTER 5

Organic Synthesis

Chapter Contents

- **Introduction**
- **Total Synthesis**
- **Methodology and Applications**
- **Total Synthesis**
- **Fatty Acid Synthesis**

Organic synthesis is the study of how we build molecules ranging from complex, biologically active natural products to new materials. Because synthesis allows a chemist to construct entirely new structures, it empowers chemists to probe the world around them in new, creative ways. Chemists and biologists have been exploring the functions of small molecules in living systems for over a century, beginning with the discovery of glucose, amino acids, vitamins, hormones, neurotransmitters, lipid mediators, and many others. Especially during the second half of the 20th century, small molecules have been used increasingly as probes ("tool compounds") of life processes. Notable early examples include the use of small-molecule neurotoxins and neurotransmitters to probe ion channels and neurotransmission, phorbol esters to probe protein kinases and signal transduction, and colchicine and cytochalasins to illuminate the molecular components of the cytoskeleton. Understanding the types of targets and processes that can be modulated with small molecules helped define the principles that underlie the rational discovery of small-molecule therapeutics

INTRODUCTION

Organic synthesis is a special branch of chemical synthesis and is concerned with the construction of organic compounds via organic reactions. Organic molecules often contain a higher level of complexity than purely inorganic compounds, so that the synthesis of organic compounds has developed into one of the most important branches of organic chemistry. There are several main areas of research within the general area of organic synthesis: *total synthesis*, *semisynthesis*, and *methodology*.

TOTAL SYNTHESIS

A total synthesis is the complete chemical synthesis of complex organic molecules from simple, commercially available (petrochemical) or natural precursors. Total synthesis may be accomplished either via a linear or convergent approach. In a *linear* synthesis—often adequate for simple structures—several steps are performed one after another until the molecule is complete; the chemical compounds made in each step are called *synthetic intermediates*. For more complex molecules, a convergent synthetic approach may be preferable, one that involves individual preparation of several "pieces" (key intermediates), which are then combined to form the desired product.

Robert Burns Woodward, who received the 1965 Nobel Prize for Chemistry for several total syntheses (e.g., his 1954 synthesis of strychnine), is regarded as the father of modern organic synthesis. Some latter-day examples include Wender›s, Holton›s, Nicolaou›s, and Danishefsky›s total syntheses of the anti-cancer therapeutic, paclitaxel (trade name, Taxol).

METHODOLOGY AND APPLICATIONS

Each step of a synthesis involves a chemical reaction, and reagents and conditions for each of these reactions must be designed to give an adequate yield of pure product, with as little work as possible. A method may already exist in the literature for making one of the early synthetic intermediates, and this method will usually be used rather than an effort to "reinvent the wheel". However, most intermediates are compounds that have never been made before, and these will normally be made using general methods developed by methodology researchers. To be useful, these methods need to give high yields, and to be reliable for a broad range of substrates. For practical applications, additional hurdles include industrial standards of safety and purity.

Methodology research usually involves three main stages: *discovery*, *optimisation*, and studies of *scope and limitations*. The *discovery* requires extensive knowledge of and experience with chemical reactivities of appropriate reagents. *Optimisation* is a process in which one or two starting compounds are tested in the reaction under a wide variety of conditions of temperature, solvent, reaction time, etc., until the optimum conditions for product yield and purity are found. Finally, the researcher tries to extend the method to a broad range of different starting materials, to find the scope and limitations. Total syntheses are sometimes used to showcase the new methodology and demonstrate its value in a real-world application. Such applications involve major industries focused especially on polymers (and plastics) and pharmaceuticals.

Stereoselective Synthesis

Most complex natural products are chiral, and the bioactivity of chiral molecules varies with the enantiomer. Historically, total syntheses targeted racemic mixtures, mixtures of both possible enantiomers, after which the racemic mixture might then be separated via chiral resolution.

In the later half of the twentieth century, chemists began to develop methods of stereoselective catalysis and kinetic resolution whereby reactions could be directed to produce only one enantiomer rather than a racemic mixture. Early examples include stereoselective hydrogenations (e.g., as reported by William Knowles and Ryōji Noyori), and functional group modifications such as the asymmetric epoxidation of Barry Sharpless; for these specific achievements, these workers were awarded the Nobel Prize in Chemistry in 2001. Such reactions gave chemists a much wider choice of enantiomerically pure molecules to start from, where previously only natural starting materials could be used. Using techniques pioneered by Robert B. Woodward and new developments in synthetic methodology, chemists became more able to take simple molecules through to more complex molecules without unwanted racemisation, by understanding stereocontrol, allowing final target molecules to be synthesised pure enantiomers (i.e., without need for resolution). Such techniques are referred to as *stereoselective synthesis*.

Synthesis Design

Elias James Corey brought a more formal approach to synthesis design, based on retrosynthetic analysis, for which he won the Nobel Prize for Chemistry in 1990. In this approach, the synthesis is planned backwards from the product, using standard rules. The steps "breaking down" the parent structure into achievable component parts are shown in a graphical scheme that uses *retrosynthetic arrows* (drawn as ⇒, which in effect, mean "is made from").

More recently, and less widely accepted, computer programs have been written for designing a synthesis based on sequences of generic "half-reactions".

TOTAL SYNTHESIS

Total synthesis is the complete chemical synthesis of a complex molecule, often a natural product, from simple, commercially available precursors. It usually refers to a process not involving the aid of biological processes, which distinguishes it from semisynthesis. The target molecules can be natural products, medicinally important active ingredients, or organic compounds of theoretical interest. Often the aim is to discover new route of synthesis for a target molecule for which there already exist known routes. Sometimes no route exists and the chemist wishes to find a viable route for the first time. One important purpose of total synthesis is the discovery of new chemical reactions and new chemical reagents.

The discovery by Friedrich Wöhler in 1828 that an organic substance, urea, can be produced from inorganic starting materials was an important conceptual milestone in chemistry as it showed for the first time that a substance previously known only as a byproduct of life could be synthesized in the laboratory.

He obtained urea artificially by treating silver cyanate with ammonium chloride. This was a rather simple synthesis.

$AgNCO + NH_4Cl \rightarrow (NH_2)_2CO + AgCl$

Haller and Blanc synthesized camphor from camphoric acid. Camphoric acid, however, had an unknown structure. In 1904, Finnish chemist Gustav Komppa became the first to succeed in manufacturing synthetic camphoric acid from diethyl oxalate and 3,3-dimethylpentanoic acid. The use of these precursors provided a demonstration for the structure of camphor, which is a complicated ring. William Perkin published another synthesis a short time later. At the time camphor was a scarce and expensive natural product with a worldwide demand. Komppa realized this and began industrial production of camphor in Tainionkoski, Finland, in 1907.

The difficulties posed by the total synthesis of a complex compound often lead to the development of new chemical reactions and routes that entail novel mechanisms, catalysts, or techniques. Finally since a total synthesis project will often span a variety of reactions, it serves to prepare chemists for pursuits in process chemistry, where encyclopedic knowledge of chemical reactions and a strong and accurate sense of chemical intuition are necessary.

Scope

A classic in total synthesis is quinine total synthesis. It has a history of many partial syntheses spanning 150 years filled with disputes and frustration. The American chemist Robert Burns Woodward was a pre-eminent figure in developing total syntheses of complex organic molecules including cholesterol, cortisone, strychnine, lysergic acid, reserpine, chlorophyll, colchicine, vitamin B_{12} and prostaglandin F-2a.

Oxytocin and vasopressin were isolated and their total synthesis reported in 1954, work for which Vincent du Vigneaud was awarded the 1955 Nobel Prize in Chemistry with the citation: "for his work on biochemically important sulphur compounds, especially for the first synthesis of a polypeptide hormone."

Another gifted chemist is Elias James Corey who won the Nobel Prize in Chemistry in 1990 for lifetime achievement in total synthesis and the development of retrosynthetic analysis.

FATTY ACID SYNTHESIS

Fatty acid synthesis is the creation of fatty acids from acetyl-CoA and NADPH through the action of enzymes called fatty acid synthases. This process takes place in the cytoplasm of the cell. Most of the acetyl-CoA which is converted into fatty acids is derived from carbohydrates via the glycolytic pathway. The glycolytic pathway also provides the glycerol with which three fatty acids can combine (by means of ester bonds) to form triglycerides (also known as "triacylglycerols", to distinguish them from fatty "acids" - or simply as "fat"), the final product of the lipogenic process. When only two fatty acids combine with glycerol and the third alcohol group is phosphorylated with a group such as phosphatidylcholine, a phospholipid is formed. Phospholipids form the bulk of the lipid bilayers that make up cell membranes and surround the organelles within the cells (e.g. the cell nucleus, mitochondria, endoplasmic reticulum, Golgi apparatus etc.)

Straight-chain Fatty Acids

Straight-chain fatty acids occur in two types: saturated and unsaturated.

Saturated straight-chain fatty acids

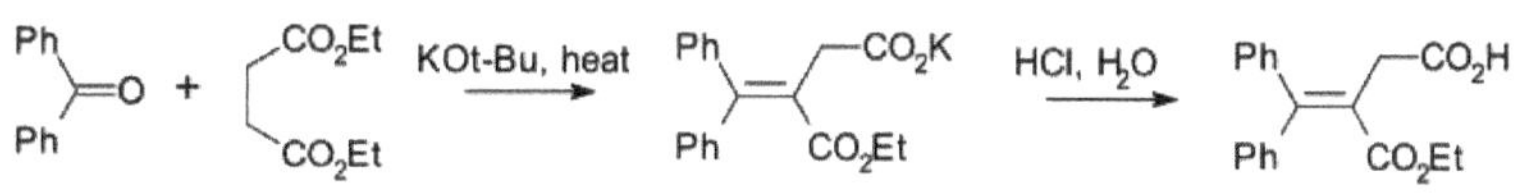

Synthesis of saturated fatty acids via Fatty Acid Synthase II in E. coli.

Much like β-oxidation, straight-chain fatty acid synthesis occurs via the six recurring reactions shown below, until the 16-carbon palmitic acid is produced.

The diagrams presented show how fatty acids are synthesized in microorganisms and list the enzymes found in Escherichia coli. These reactions are performed by fatty acid synthase II (FASII), which in general contain multiple enzymes that act as one complex. FASII is present in prokaryotes, plants, fungi, and parasites, as well as in mitochondria.

In animals, as well as some fungi such as yeast, these same reactions occur on fatty acid synthase I (FASI), a large dimeric protein that has all of the enzymatic activities required to create a fatty acid. FASI is less efficient than FASII; however, it allows for the formation of more molecules, including medium-chain fatty acids via early chain termination.

Once a 16:0 carbon fatty acid has been formed, it can undergo a number of modifications, resulting in desaturation and/or elongation. Elongation, starting with stearate (18:0), is performed mainly in the ER by several membrane-bound enzymes. The enzymatic steps involved in the elongation process are principally the same as those carried out by FAS, but the four principal successive steps of the elongation are performed by individual proteins, which may be physically associated.

Step	Enzyme	Reaction	Description
(a)	Acetyl CoA:ACP transacylase	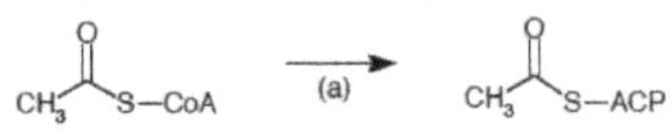	Activates acetyl CoA for reaction with malonyl-ACP
(b)	Malonyl CoA:ACP transacylase	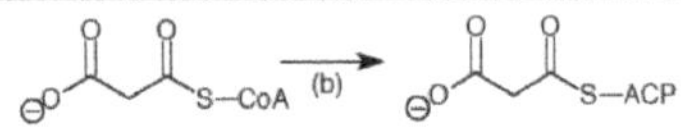	Activates malonyl CoA for reaction with acetyl-ACP
(c)	3-ketoacyl-ACP synthase	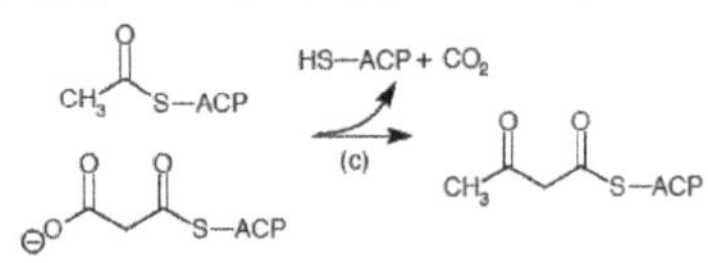	Reacts priming acetyl-ACP with chain-extending malonyl-ACP.
(d)	3-ketoacyl-ACP reductase	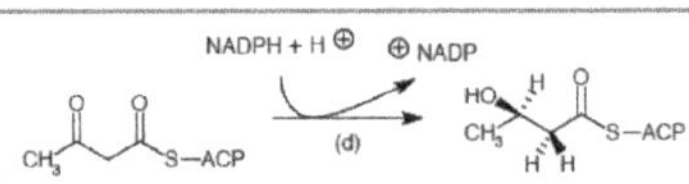	Reduces the carbon 3 ketone to a hydroxyl group
(e)	3-Hydroxyacyl ACP dehydrase	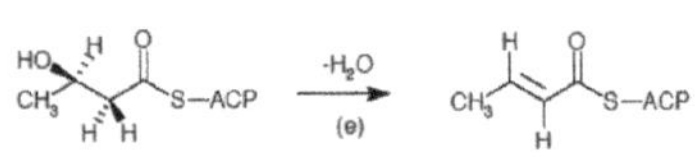	Removes water
(f)	Enoyl-ACP reductase	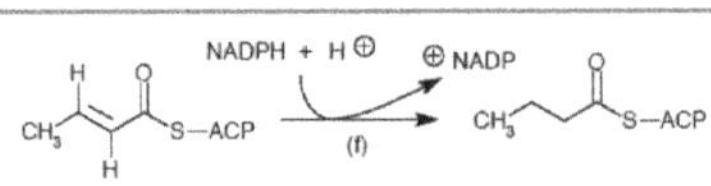	Reduces the C2-C3 double bond.

Abbreviations: ACP – Acyl carrier protein, CoA – Coenzyme A, NADP – Nicotinamide adenine dinucleotide phosphate.

Note that during fatty synthesis the reducing agent is NADPH, whereas NAD is the oxidizing agent in beta-oxidation (the breakdown of fatty acids to acetyl-CoA). This difference exemplifies a general principle that NADPH is consumed during biosynthetic reactions, whereas NADH is generated in energy-yielding reactions. (Thus NADPH is also required for the synthesis of cholesterol from acetyl-CoA; while NADH is generated during glycolysis.) The source of the NADPH is two-fold. When malate is oxidatively decarboxylated by "$NADP^+$-linked malic enzyme" pyruvate, CO_2 and NADPH are formed. NADPH is also formed by the pentose phosphate pathway which converts glucose into ribose, which can be used in synthesis of nucleotides and nucleic acids, or it can be catabolized to pyruvate.

Glycolytic end Products are used in the Conversion of Carbohydrates into Fatty Acids

In humans, fatty acids are formed from carbohydrates predominantly in the liver and adipose tissue, as well as in the mammary glands during lactation.

The pyruvate produced by glycolysis is an important intermediary in the conversion of carbohydrates into fatty acids and cholesterol. This occurs via the conversion of pyruvate into acetyl-CoA in the mitochondrion. However, this acetyl CoA needs to be transported into cytosol where the synthesis of fatty acids and cholesterol occurs. This cannot occur directly. To obtain cytosolic acetyl-CoA, citrate (produced by the condensation of acetyl CoA with oxaloacetate) is removed from the citric acid cycle and carried across the inner mitochondrial membrane into the cytosol. There it is cleaved by ATP citrate lyase into acetyl-CoA and oxaloacetate. The oxaloacetate can be used for gluconeogenesis (in the liver), or it can be returned into mitochondrion as malate. The cytosolic acetyl-CoA is carboxylated by acetyl CoA carboxylase into malonyl CoA, the first committed step in the synthesis of fatty acids.

Animals cannot Resynthesize Carbohydrates from Fatty Acids

The main fuel stored in the bodies of animals is fat. The young adult human's fat stores average between about 10-20 kg, but varies greatly depending on age, gender, and individual disposition. By contrast the human body stores only about 400 g of glycogen, of which 300 g is locked inside the skeletal muscles and is unavailable to the body as a whole. The 100 g or so of glycogen stored in the liver is depleted within one day of starvation. Thereafter the glucose that is released into the blood by the liver for general use by the body tissues, has to be synthesized from the glucogenic amino acids and a few other gluconeogenic substrates, which do not include fatty acids.

Fatty acids are broken down to acetyl-CoA by means of beta oxidation inside the mitochondria, whereas fatty acids are synthesized from acetyl-CoA outside the mitochondrion, in the cytosol. The two pathways are distinct, not only in where they occur, but also in the reactions that occur, and the substrates that are used. The two pathways are mutually inhibitory, preventing the acetyl-CoA produced by beta-oxidation from entering the synthetic pathway via the acetyl-CoA carboxylase reaction. It can also not be converted to pyruvate as the pyruvate decarboxylation reaction is irreversible. Instead it condenses with oxaloacetate, to enter the citric acid cycle. During each turn of the cycle, two carbon atoms leave the cycle as CO_2 in the decarboxylation reactions catalyzed by isocitrate dehydrogenase and alpha-ketoglutarate dehydrogenase. Thus each turn of the citric acid cycle oxidizes an acetyl-CoA unit while regenerating the oxaloacetate molecule with which the acetyl-CoA had originally combined to form citric acid. The decarboxylation reactions occur before malate is formed in the cycle. This is the only substance that can be removed from the mitochondrion to enter the gluconeogenic pathway to form glucose or glycogen in the liver or any other tissue. There can therefore be no net conversion of fatty acids into glucose. Only plants possess the enzymes to convert acetyl-CoA into oxaloacetate from which malate can be formed to ultimately be converted to glucose.

Regulation

Acetyl-CoA is formed into malonyl-CoA by acetyl-CoA carboxylase, at which point malonyl-CoA is destined to feed into the fatty acid synthesis pathway. Acetyl-CoA carboxylase is the point of regulation in saturated straight-chain fatty acid synthesis, and is subject to both phosphorylation and allosteric regulation. Regulation by phosphorylation occurs mostly in mammals, while allosteric regulation occurs in most organisms. Allosteric control occurs as feedback inhibition by palmitoyl-CoA and activation by citrate. When there are high levels of palmitoyl-CoA, the final product of saturated fatty acid synthesis, it allosterically inactivates acetyl-CoA carboxylase to prevent a build-up of fatty acids in cells. Citrate acts to activate acetyl-CoA carboxylase under high levels, because high levels indicate that there is enough acetyl-CoA to feed into the Krebs cycle and produce energy.

High plasma levels of insulin in the blood plasma (e.g. after meals) cause the dephosphorylation of acetyl-CoA carboxylase, thus promoting the formation of malonyl-CoA from acetyl-CoA, and consequently the conversion of carbohydrates into fatty acids, while epinephrine and glucagon (released into the blood during starvation and exercise) cause the phosphorylation of this enzyme, inhibiting lipogenesis in favor of fatty acid oxidation via beta-oxidation.

Anaerobic Desaturation

Many bacteria use the anaerobic pathway for synthesizing unsaturated fatty acids. This pathway does not utilize oxygen and is dependent on enzymes to insert the double bond before elongation utilizing the normal fatty acid synthesis machinery. In *Escherichia coli*, this pathway is well understood.

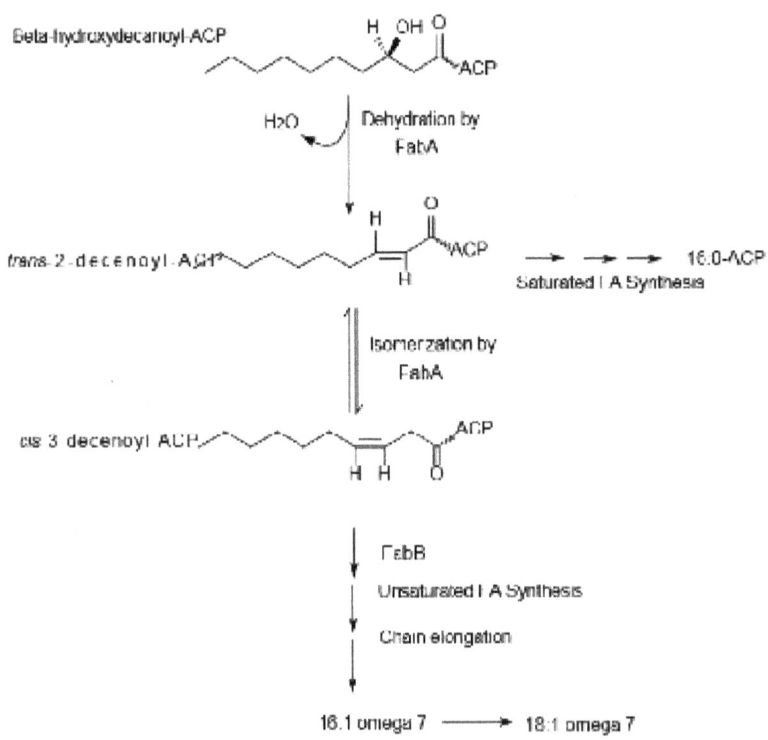

Synthesis of unsaturated fatty acids via anaerobic desaturation.

- FabA is a β-hydroxydecanoyl-ACP dehydrase – it is specific for the 10-carbon saturated fatty acid synthesis intermediate (β-hydroxydecanoyl-ACP).

- FabA catalyzes the dehydration of β-hydroxydecanoyl-ACP, causing the release of water and insertion of the double bond between C7 and C8 counting from the methyl end. This creates the trans-2-decenoyl intermediate.
- Either the trans-2-decenoyl intermediate can be shunted to the normal saturated fatty acid synthesis pathway by FabB, where the double bond will be hydrolyzed and the final product will be a saturated fatty acid, or FabA will catalyze the isomerization into the cis-3-decenoyl intermediate.
- FabB is a β-ketoacyl-ACP synthase that elongates and channels intermediates into the mainstream fatty acid synthesis pathway. When FabB reacts with the cis-decenoyl intermediate, the final product after elongation will be an unsaturated fatty acid.
- The two main unsaturated fatty acids made are Palmitoleoyl-ACP (16:1ω7) and cis-vaccenoyl-ACP (18:1ω7).

Most bacteria that undergo anaerobic desaturation contain homologues of FabA and FabB. Clostridia are the main exception; they have a novel enzyme, yet to be identified, that catalyzes the formation of the cis double bond.

Regulation

This pathway undergoes transcriptional regulation by FadR and FabR. FadR is the more extensively studied protein and has been attributed bifunctional characteristics. It acts as an activator of *fabA* and *fabB* transcription and as a repressor for the β-oxidation regulon. In contrast, FabR acts as a repressor for the transcription of fabA and fabB.

Aerobic Desaturation

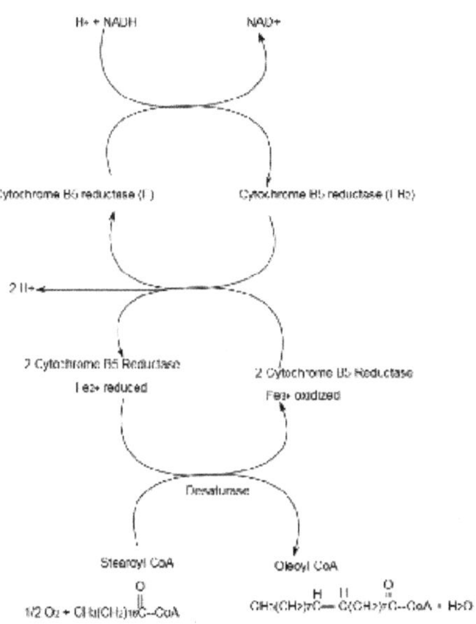

Synthesis of unsaturated fatty acids via aerobic desaturation.

Aerobic desaturation is the most widespread pathway for the synthesis of unsaturated fatty acids. It is utilized in all eukaryotes and some prokaryotes. This pathway utilizes desaturases to synthesize unsaturated fatty acids from full-length saturated fatty acid substrates. All desaturases require oxygen and ultimately consume NADH even though desaturation is an oxidative process. Desaturases are specific for the double bond they induce in the substrate. In Bacillus subtilis, the desaturase, Δ^5-Des, is specific for inducing a cis-double bond at the Δ^5 position. *Saccharomyces cerevisiae* contains one desaturase, Ole1p, which induces the cis-double bond at Δ^9.

In mammals the aerobic desaturation is catalyzed by a complex of three membrane-bound enzymes (*NADH-cytochrome b_5 reductase, cytochrome b_5*, and a *desaturase*). These enzymes allow molecular oxygen, O_2, to interact with the saturated fatty acyl-CoA chain, forming a double bond and two molecules of water, H_2O. Two electrons come from NADH + H^+ and two from the single bond in the fatty acid chain. These mammalian enzymes are, however, incapable of introducing double bonds at carbon atoms beyond C-9 in the fatty acid chain.(For the numbering of the carbon atoms in a fatty acid.) Hence mammals cannot synthesize linoleate or linolenate (which have double bonds at the C-12 (= Δ^{12}), or the C-12 and C-15 (= $\Delta^{12}$and Δ^{15}) positions, respectively, as well as at the Δ^9 position), nor the polyunsaturated, 20-carbon arachidonic acid that is derived from linoleate. These are all termed essential fatty acids, meaning that they are required by the organism, but can only be supplied via the diet. (Arachidonic acid is the precursor the prostaglandins which fulfill a wide variety of functions as local hormones.)

Regulation

In *B. subtilis*, this pathway is regulated by a two-component system: DesK and DesR. DesK is a membrane-associated kinase and DesR is a transcriptional regulator of the *des* gene. The regulation responds to temperature; when there is a drop in temperature, this gene is upregulated. Unsaturated fatty acids increase the fluidity of the membrane and stabilize it under lower temperatures. DesK is the sensor protein that, when there is a decrease in temperature, will autophosphorylate. DesK-P will transfer its phosphoryl group to DesR. Two DesR-P proteins will dimerize and bind to the DNA promoters of the *des* gene and recruit RNA polymerase to begin transcription.

Pseudomonas aeruginosa

In general, both anaerobic and aerobic unsaturated fatty acid synthesis will not occur within the same system, however *Pseudomonas aeruginosa* and *Vibrio* ABE-1 are exceptions. While *P. aeruginosa* undergoes primarily anaerobic desaturation, it also undergoes two aerobic pathways. One pathway utilizes a Δ^9-desaturase (DesA) that catalyzes a double bond formation in membrane lipids. Another pathway uses two proteins, DesC and DesB, together to act as a Δ^9-desaturase, which inserts a double bond into a saturated fatty acid-CoA molecule. This second pathway is regulated by repressor protein DesT. DesT is also a repressor of *fabAB* expression for anaerobic desaturation when in presence of exogenous unsaturated fatty acids. This functions to coordinate the expression of the two pathways within the organism.

Branched-chain Fatty Acids

Branched-chain fatty acids are usually saturated and are found in two distinct families: the iso-series and anteiso-series. It has been found that Actinomycetales contain unique branch-chain fatty acid synthesis mechanisms, including that which forms tuberculosteric acid.

Branch-chain Fatty Acid Synthesizing System

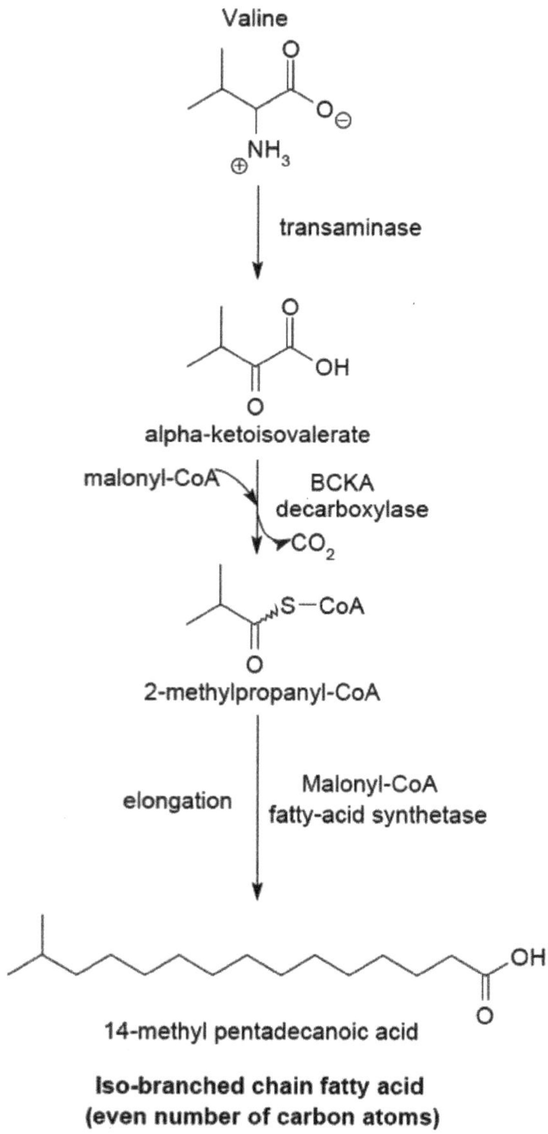

Valine primer

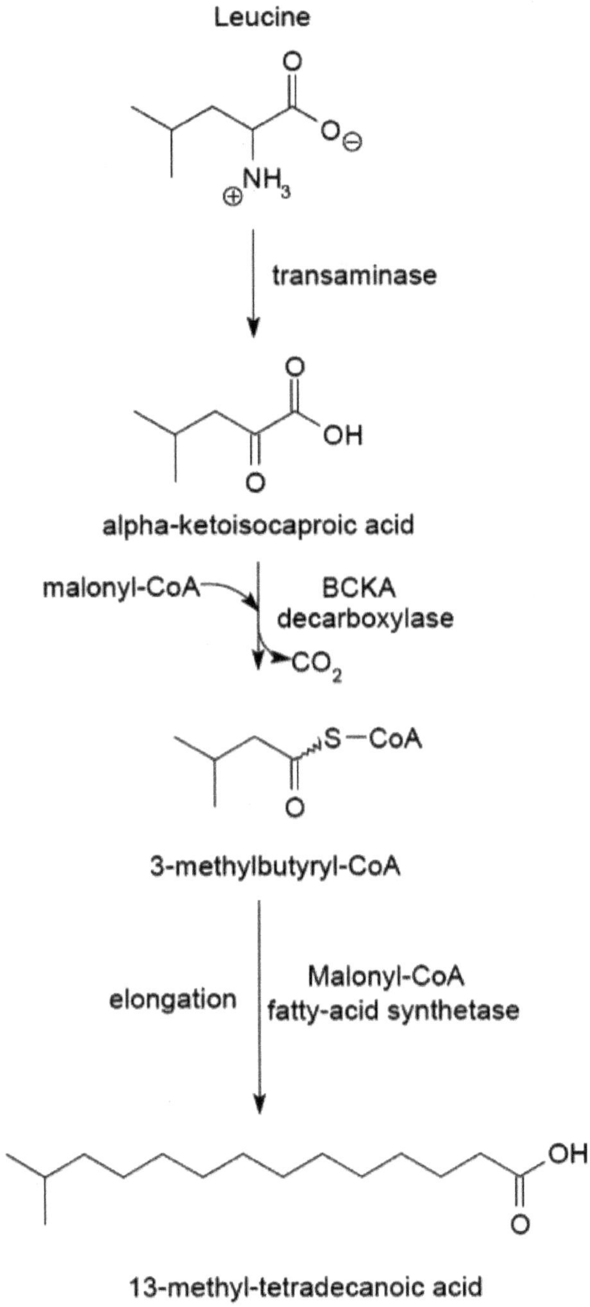

Iso-branched chain fatty acid
(odd number of carbon atoms)

Leucine Primer

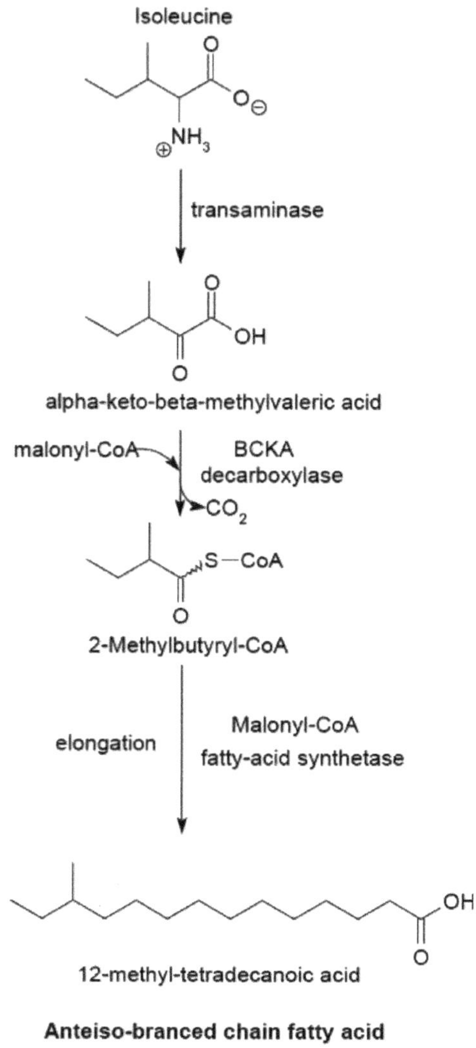

Isoleucine primer

Synthetic pathways of the branched-chain fatty acid synthesizing system given differing primers.

The branched-chain fatty acid synthesizing system uses α-keto acids as primers. This system is distinct from the branched-chain fatty acid synthetase that utilizes short-chain acyl-CoA esters as primers. α-Keto acid primers are derived from the transamination and decarboxylation of valine, leucine, and isoleucine to form 2-methylpropanyl-CoA, 3-methylbutyryl-CoA, and 2-Methylbutyryl-CoA, respectively.2-Methylpropanyl-CoA primers derived from valine are elongated to produce even-numbered iso-series fatty acids such as 14-methyl-pentadecanoic (isopalmitic) acid, and 3-methylbutyryl-CoA primers from leucine may be used to form odd-numbered iso-series fatty acids such as

13-methyl-tetradecanoic acid. 2-Methylbutyryl-CoA primers from isoleucine are elongated to form anteiso-series fatty acids containing an odd number of carbon atoms such as 12-Methyl tetradecanoic acid. Decarboxylation of the primer precursors occurs through the branched-chain α-keto acid decarboxylase (BCKA) enzyme. Elongation of the fatty acid follows the same biosynthetic pathway in *Escherichia coli* used to produce straight-chain fatty acids where malonyl-CoA is used as a chain extender. The major end products are 12–17 carbon branched-chain fatty acids and their composition tends to be uniform and characteristic for many bacterial species.

BCKA Decarboxylase and Relative Activities of α-keto Acid Substrates

The BCKA decarboxylase enzyme is composed of two subunits in a tetrameric structure (A_2B_2) and is essential for the synthesis of branched-chain fatty acids. It is responsible for the decarboxylation of α-keto acids formed by the transamination of valine, leucine, and isoleucine and produces the primers used for branched-chain fatty acid synthesis. The activity of this enzyme is much higher with branched-chain α-keto acid substrates than with straight-chain substrates, and in Bacillus species its specificity is highest for the isoleucine-derived α-keto-β-methylvaleric acid, followed by α-ketoisocaproate and α-ketoisovalerate. The enzyme's high affinity toward branched-chain α-keto acids allows it to function as the primer donating system for branched-chain fatty acid synthetase.

Substrate	BCKA activity	CO2 Produced (nmol/min mg)	Km (μM)	Vmax (nmol/min mg)
$_L$-α-keto-β-methyl-valerate	100%	19.7	<1	17.8
α-Ketoisovalerate	63%	12.4	<1	13.3
α-Ketoisocaproate	38%	7.4	<1	5.6
Pyruvate	25%	4.9	51.1	15.2

Factors Affecting Chain Length and Pattern Distribution

α-Keto acid primers are used to produce branched-chain fatty acids that, in general, are between 12 and 17 carbons in length. The proportions of these branched-chain fatty acids tend to be uniform and consistent among a particular bacterial species but may be altered due to changes in malonyl-CoA concentration, temperature, or heat-stable factors (HSF) present. All of these factors may affect chain length, and HSFs have been demonstrated to alter the specificity of BCKA decarboxylase for a particular α-keto acid substrate, thus shifting the ratio of branched-chain fatty acids produced. An increase in malonyl-CoA concentration has been shown to result in a larger proportion of C17 fatty acids produced, up until the optimal concentration (≈20μM) of malonyl-CoA is reached.

Decreased temperatures also tend to shift the fatty-acid distribution slightly toward C17 fatty-acids in *Bacillus* species.

Branch-chain fatty Acid Synthase

This system functions similarly to the branch-chain fatty acid synthesizing system, however it uses short-chain carboxylic acids as primers instead of alpha-keto acids. In general, this method is used by bacteria that do not have the ability to perform the branch-chain fatty acid system using alpha-keto primers. Typical short-chain primers include isovalerate, isobutyrate, and 2-methyl butyrate. In general, the acids needed for these primers are taken up from the environment; this is often seen in ruminal bacteria.

The overall reaction is:

Isobutyryl-CoA + 6 malonyl-CoA +12 NADPH + $12H^+$ → Isopalmitic acid + 6 CO_2 12 NADP + 5 H_2O + 7 CoA

The difference between (straight-chain) fatty acid synthase and branch-chain fatty acid synthase is substrate specificity of the enzyme that catalyzes the reaction of acyl-CoA to acyl-ACP.

Omega-Alicyclic Fatty Acids

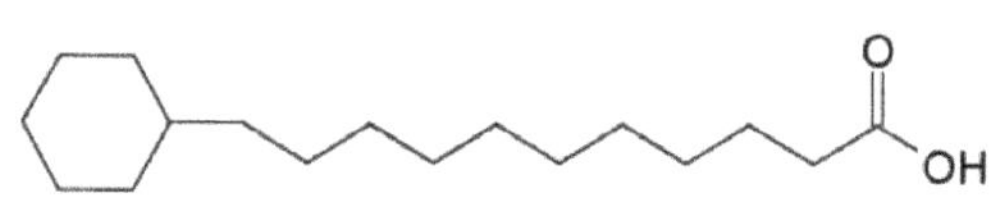

11-cyclohexylundecanoic acid

Omega-alicyclic fatty acid

Omega-alicyclic fatty acids typically contain an omega-terminal propyl or butyryl cyclic group and are some of the major membrane fatty acids found in several species of bacteria. The fatty acid synthetase used to produce omega-alicyclic fatty acids is also used to produce membrane branched-chain fatty acids. In bacteria with membranes composed mainly of omega-alicyclic fatty acids, the supply of cyclic carboxylic acid-CoA esters is much greater than that of branched-chain primers. The synthesis of cyclic primers is not well understood but it has been suggested that mechanism involves the conversion of sugars to shikimic acid which is then converted to cyclohexylcarboxylic acid-CoA esters that serve as primers for omega-alicyclic fatty acid synthesis

Tuberculostearic acid Synthesis

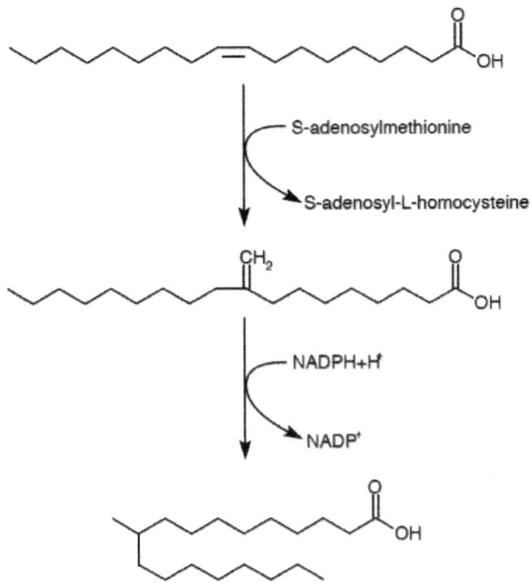

Mechanism of the synthesis of Tuberculostearic acid.

Tuberculostearic acid ($_D$-10-Methylstearic acid) is a saturated fatty acid that is known to be produced by Mycobacterium spp. and two species of *Streptomyces*. It is formed from the precursor oleic acid (a monosaturated fatty acid). After oleic acid is esterified to a phospholipid, S-adenosyl-methionine donates a methyl group to the double bond of oleic acid. This methylation reaction forms the intermediate 10-methylene-octadecanoyal. Successive reduction of the residue, with NADPH as a cofactor, results in 10-methylstearic acid

REFERECES

1. Dijkstra, Albert J., R. J. Hamilton, and Wolf Hamm. "Fatty Acid Biosynthesis." Trans Fatty Acids. Oxford: Blackwell Pub., 2008. 12. Print.
2. MetaCyc pathway: superpathway of fatty acids biosynthesis (E. coli) .
3. "Fatty Acids: Straight-chain Saturated, Structure, Occurrence and Biosynthesis." Lipid Library – Lipid Chemistry, Biology, Technology and Analysis. Web. 30 Apr. 2011. <http://lipidlibrary.aocs.org/lipids/fa_sat/index.htm>.
4. MetaCyc pathway: stearate biosynthesis I (animals) .
5. MetaCyc pathway: very long chain fatty acid biosynthesis II .
6. Stryer, Lubert (1995). Biochemistry. (Fourth ed.). New York: W.H. Freeman and Company. pp. 559–565, 614–623. ISBN 0 7167 2009 4.

7. Ferre, P.; F. Foufelle (2007). SREBP-1c Transcription Factor and Lipid Homeostasis: Clinical Perspective . Hormone Research. **68** *(2): 72–82. doi:10.1159/000100426. PMID 17344645. Retrieved 2010-08-30. this process is outlined graphically in page 73*

8. Voet, Donald; Judith G. Voet; Charlotte W. Pratt (2006). Fundamentals of Biochemistry, 2nd Edition. John Wiley and Sons, Inc. pp. 547, 556. ISBN 0-471-21495-7.

9. Sloan, A.W; Koeslag, J.H.; Bredell, G.A.G. (1973). Body composition work capacity and work efficiency of active and inactive young men . European Journal of Applied Physiology. **32**: 17–24. doi:10.1007/bf00422426.

10. Stryer, Lubert (1995). Biochemistry. (Fourth ed.). New York: W.H. Freeman and Company. pp. 581–602, 613, 775–778. ISBN 0 7167 2009 4.

11. Stryer, Lubert (1995). "Fatty acid metabolism.". In: Biochemistry. (Fourth ed.). New York: W.H. Freeman and Company. pp. 603–628. ISBN 0 7167 2009 4.

12. Diwan, Joyce J. "Fatty Acid Synthesis." Rensselaer Polytechnic Institute (RPI) :: Architecture, Business, Engineering, IT, Humanities, Science. Web. 30 Apr. 2011. <http://rpi.edu/dept/bcbp/molbiochem/MBWeb/mb2/part1/fasynthesis.htm>.

13. Feng, Youjun, and John ECronan. "Complex binding of the FabR repressor of bacterial unsaturated fatty acid biosynthesis to its cognate promoters." Molecular microbiology 80.1 (2011):195–218.

14. Zhu, Lei, et al. Functions of the Clostridium acetobutylicium FabF and FabZ proteins in unsaturated fatty acid biosynthesis. BMC microbiology 9(2009):119.

15. Wang, Haihong, and John ECronan. "Functional replacement of the FabA and FabB proteins of Escherichia coli fatty acid synthesis by Enterococcus faecalis FabZ and FabF homologues. Journal of biological chemistry 279.33 (2004):34489-95.

16. Mansilla, Mara C, and Diegode Mendoza. The Bacillus subtilis desaturase: a model to understand phospholipid modification and temperature sensing. Archives of microbiology 183.4 (2005):229-35.

17. Wada, M, N. Fukunaga, and S. Sasaki. "Mechanism of biosynthesis of unsaturated fatty acids in Pseudomonas sp. strain E-3, a psychrotrophic bacterium." Journal of bacteriology 171.8 (1989):4267-71.

18. Subramanian, Chitra, Charles ORock, and Yong-MeiZhang. "DesT coordinates the expression of anaerobic and aerobic pathways for unsaturated fatty acid biosynthesis in Pseudomonas aeruginosa." Journal of bacteriology 192.1 (2010):280-5.

19. Morita, N, et al. "Both the anaerobic pathway and aerobic desaturation are involved in the synthesis of unsaturated fatty acids in Vibrio sp. strain ABE-1." FEBS letters 297.1–2 (1992):9–12.

20. Zhu, Kun, et al. "Two aerobic pathways for the formation of unsaturated fatty acids in Pseudomonas aeruginosa." Molecular microbiology 60.2 (2006):260-73.

21. Kaneda, Toshi. Iso- and Anteiso-Fatty Acids in Bacteria: Biosynthesis, Function, and Taxonomic Significance. Microbiological Reviews 55.2 (1991): 288–302

22. "Branched-chain Fatty Acids, Phytanic Acid, Tuberculostearic Acid Iso/anteiso-Fatty Acids. Lipid Library – Lipid Chemistry, Biology, Technology and Analysis. Web. 1 May 2011. http://lipidlibrary.aocs.org/lipids/fa_branc/index.htm.

23. Naik, Devaray N., and Toshi Kaneda. Biosynthesis of Branched Long-chain Fatty Acids by Species of Bacillus: Relative Activity of Three α-keto Acid Substrates and Factors Affecting Chain Length. Can. J. Microbiol. 20 (1974): 1701–708.

24. Oku, Hirosuke, and Toshi Kaneda. Biosynthesis of Branched-chain Fatty Acids in Bacillis Subtilis. The Journal of Biological Chemistry 263.34 (1988): 18386-8396.

25. Christie, William W. "Fatty Acids: Natural Alicyclic Structures, Occurrence, and Biochemistry." The AOCS Lipid Library. 5 Apr. 2011. Web. 24 Apr. 2011. <http://lipidlibrary.aocs.org/lipids/fa_cycl/file.pdf>.

26. Ratledge, Colin, and John Stanford. The Biology of the Mycobacteria. London: Academic, 1982. Print.

27. Kubica, George P., and Lawrence G. Wayne. The Mycobacteria: a Sourcebook. New York: Dekker, 1984.

CHAPTER 6

Theory of Aromaticity

Chapter Contents

Aromatic compounds comprise a class of hydrocarbons that include a six-member, unsaturated carbon ring in which the pi bond valence electrons are completely delocalized or conjugated. These compounds are stable and abundant in both natural and synthetic forms. The simplest of the aromatic compounds is benzene (C_6H_6), a flammable carcinogen, yet an industrially important chemical. The name aromatic is based on the strong aromas of many of the larger aromatic compounds. Diamonds and graphite, while not considered aromatic compounds, demonstrate delocalized electron sharing over very long atomic distances.

INTRODUCTION

In organic chemistry, the term **aromaticity** is used to describe a cyclic (ring-shaped), planar (flat) molecule with a ring of resonance bonds that exhibits more stability than other geometric or connective arrangements with the same set of atoms. Aromatic molecules are very stable, and do not break apart easily to react with other substances. Organic compounds that are not aromatic are classified as aliphatic compounds—they might be cyclic, but only aromatic rings have special stability (low reactivity).

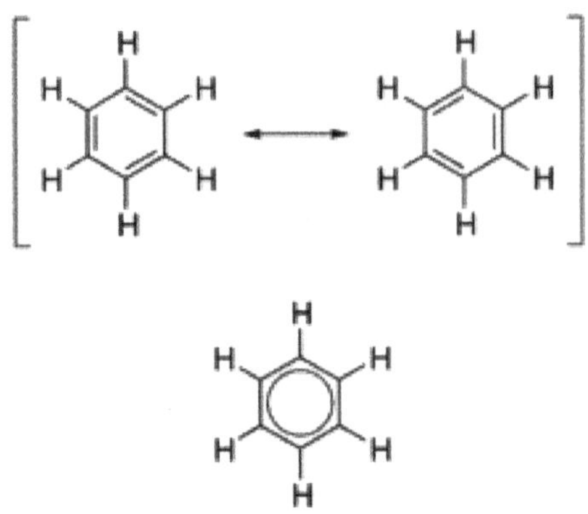

Two different resonance forms of benzene (top) combine to produce an average structure (bottom).

Since the most common aromatic compounds are derivatives of benzene (an aromatic hydrocarbon common in petroleum and its distillates), the word "aromatic" occasionally refers informally to benzene derivatives, and so it was first defined. Nevertheless, many non-benzene aromatic compounds exist. In living organisms, for example, the most common aromatic rings are the double-ringed bases in RNA and DNA. An aromatic functional group or other substituent is called an aryl group.

The earliest use of the term aromatic was in an article by August Wilhelm Hofmann in 1855. Hofmann used the term for a class of benzene compounds, many of which have odors (aromas), unlike pure saturated hydrocarbons. Aromaticity as a chemical property bears no general relationship with the olfactory properties of such compounds (how they smell), although in 1855, before the structure of benzene or organic compounds was understood, chemists like Hofmann were beginning to understand that odiferous molecules from plants, such as terpenes, had chemical properties that we recognize today are similar to unsaturated petroleum hydrocarbons like benzene.

In terms of the electronic nature of the molecule, aromaticity describes a conjugated system often made of alternating single and double bonds in a ring. This configuration allows for the electrons in the molecule›s pi system to be delocalized around the ring, increasing the molecule›s stability. The molecule cannot be represented by one structure, but rather a resonance hybrid of different structures, such as with the two resonance structures of benzene. These molecules cannot be found in either one of these representations, with the longer single bonds in one location and the shorter double bond in another (see Theory below). Rather, the molecule exhibits bond lengths in between those of single and double bonds. This commonly seen model of aromatic rings, namely the idea that benzene was formed from a six-membered carbon ring with alternating single and double bonds (cyclohexatriene), was developed by August Kekulé (see History below). The model for benzene consists of two resonance forms, which corresponds to the double and single bonds superimposing to produce six one-and-a-half bonds. Benzene is a more stable molecule than would be expected without accounting for charge delocalization.

Theory

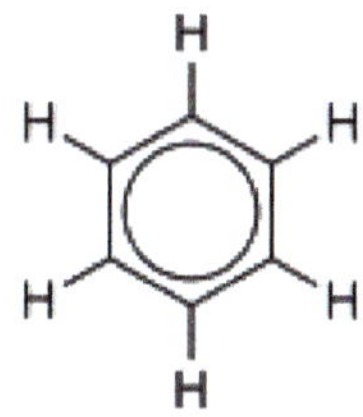

As is standard for resonance diagrams, the use of a double-headed arrow indicates that two structures are not distinct entities but merely hypothetical possibilities. Neither is an accurate representation of the *actual* compound, which is best represented by a hybrid (average) of these structures, as can be seen at right. A C=C bond is shorter than a C−C bond, but benzene is perfectly hexagonal—all six carbon–carbon bonds have the same length, intermediate between that of a single and that of a double bond.

In a cyclic molecule with three alternating double bonds, cyclohexatriene, the bond length of the single bond would be 1.54 Å and that of the double bond would be 1.34 Å. However, in a molecule of benzene, the length of each of the bonds is 1.40 Å, between these two lengths.

A better representation is that of the circular π-bond (Armstrong's *inner cycle*), in which the electron density is evenly distributed through a π-bond above and below the ring. This model more correctly represents the location of electron density within the aromatic ring.

The single bonds are formed from overlap of hybridized atomic sp^2-orbitals in line between the carbon nuclei—these are called σ-bonds. Double bonds consist of a σ-bond and a π-bond. The π-bonds are formed from overlap of atomic p-orbitals above and below the plane of the ring. The following diagram shows the positions of these p-orbitals:

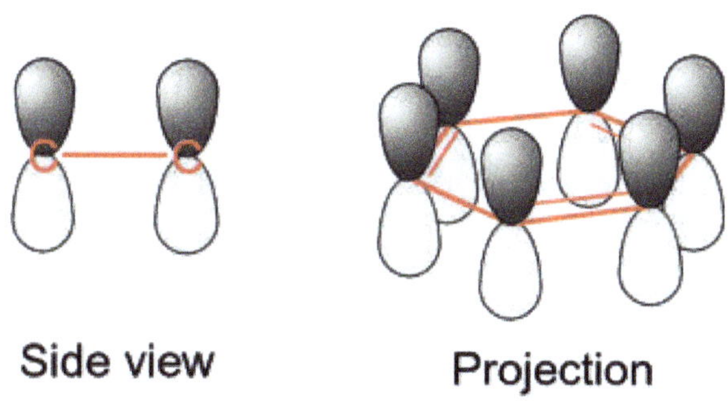

Since they are out of the plane of the atoms, these orbitals can interact with each other freely, and become delocalized. This means that, instead of being tied to one atom of carbon, each electron is shared by all six in the ring. Thus, there are not enough electrons to form double bonds on all the carbon atoms, but the "extra" electrons strengthen all of the bonds on the ring equally. The resulting molecular orbital is considered to have ϖ *symmetry*.

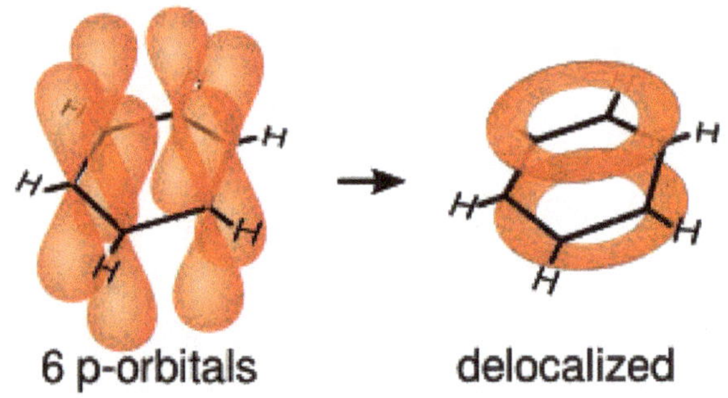

The Term "Aromatic"

The first known use of the word "aromatic" as a *chemical* term—namely, to apply to compounds that contain the phenyl radical—occurs in an article by August Wilhelm Hofmann in 1855. If this is indeed the earliest introduction of the term, it is curious that Hofmann says nothing about why he introduced an adjective indicating olfactory character to apply to a group of chemical substances only some of which have notable aromas. Also, many of the most odoriferous organic substances known are terpenes, which are not aromatic in the chemical sense. But terpenes and benzenoid substances do have a chemical characteristic in common, namely higher unsaturation than many aliphatic compounds, and Hofmann may not have been making a distinction between the two categories. Many of the earliest-known examples of aromatic compounds, such as benzene and toluene, have distinctive pleasant smells. This property led to the term aromatic for this class of compounds, and hence the term aromaticity for the eventually discovered electronic property.

The Structure of the Benzene Ring

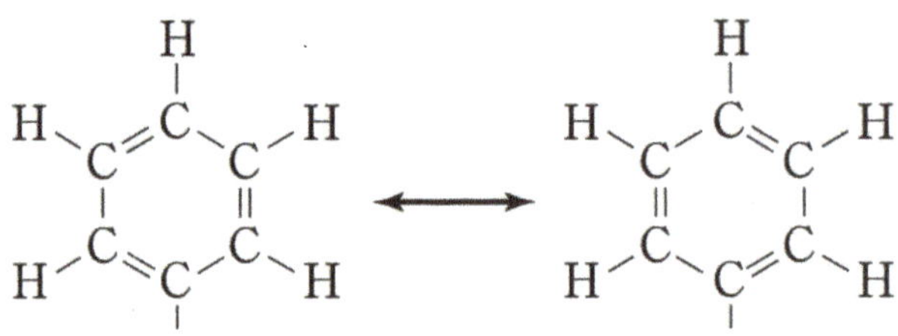

Historic benzene formulae as proposed by August Kekulé in 1865.

In the 19th century chemists found it puzzling that benzene could be so unreactive toward addition reactions, given its presumed high degree of unsaturation. The cyclohexatriene structure for benzene was first proposed by August Kekulé in 1865. Most chemists were quick to accept this structure, since it accounted for most of the known isomeric relationships of aromatic chemistry. The hexagonal structure explains why only one isomer of benzene exists and why disubstituted compounds have three isomers.

Between 1897 and 1906, J. J. Thomson, the discoverer of the electron, proposed three equivalent electrons between each pair of carbon atoms in benzene. An explanation for the exceptional stability of benzene is conventionally attributed to Sir Robert Robinson, who was apparently the first (in 1925) to coin the term *aromatic sextet* as a group of six electrons that resists disruption.

In fact, this concept can be traced further back, via Ernest Crocker in 1922, to Henry Edward Armstrong, who in 1890 wrote the (six) centric affinities act within a cycle... benzene may be represented by a **double ring** (*sic*) ... and when an additive compound is formed, the inner cycle of affinity suffers disruption, the contiguous carbon-atoms to which nothing has been attached of necessity acquire the ethylenic condition".

Here, Armstrong is describing at least four modern concepts. First, his "affinity" is better known nowadays as the electron, which was to be discovered only seven years later by J. J. Thomson. Second, he is describing electrophilic aromatic substitution, proceeding (third) through a Wheland intermediate, in which (fourth) the conjugation of the ring is broken. He introduced the symbol **C** centered on the ring as a shorthand for the *inner cycle*, thus anticipating Erich Clar's notation. It is argued that he also anticipated the nature of wave mechanics, since he recognized that his affinities had direction, not merely being point particles, and collectively having a distribution that could be altered by introducing substituents onto the benzene ring (much as the distribution of the electric charge in a body is altered by bringing it near to another body).

The quantum mechanical origins of this stability, or aromaticity, were first modelled by Hückel in 1931. He was the first to separate the bonding electrons into sigma and pi electrons.

Aromaticity of an arbitrary aromatic compound can be measured quantitatively by the nucleus-independent chemical shift (NICS) computational method and aromaticity percentage methods.

Characteristics of Aromatic (aryl) Compounds

An aromatic (or aryl) compound contains a set of covalently bound atoms with specific characteristics:

- A delocalized conjugated π system, most commonly an arrangement of alternating single and double bonds
- Coplanar structure, with all the contributing atoms in the same plane
- Contributing atoms arranged in one or more rings
- A number of π delocalized electrons that is even, but not a multiple of 4. That is, $4n + 2$ π-electrons, where $n = 0, 1, 2, 3$, and so on. This is known as Hückel›s rule.

According to Hückel›s rule, if a molecule has $4n + 2$ π-electrons, it is aromatic, but if it has $4n$ π-electrons and has characteristics 1–3 above, the molecule is said to be antiaromatic. Whereas benzene is aromatic (6 electrons, from 3 double bonds), cyclobutadiene is antiaromatic, since the number of π delocalized electrons is 4, which of course is a multiple of 4. The cyclobutadienide(2−) ion, however, is aromatic (6 electrons). An atom in an aromatic system can have other electrons that are not part of the system, and are therefore ignored for the $4n + 2$ rule. In furan, the oxygen atom is sp^2 hybridized. One lone pair is in the π system and the other in the plane of the ring (analogous to the C–H bond in the other positions). There are 6 π-electrons, so furan is aromatic.

Aromatic molecules typically display enhanced chemical stability, compared to similar non-aromatic molecules. A molecule that can be aromatic will tend to change toward aromaticity, and the added stability changes the chemistry of the molecule. Aromatic compounds undergo electrophilic aromatic substitution and nucleophilic aromatic substitution reactions, but not electrophilic addition reactions as happens with carbon–carbon double bonds.

The circulating π-electrons in an aromatic molecule produce ring currents that oppose the applied magnetic field in NMR.The NMR signal of protons in the plane of an aromatic ring are shifted substantially further down-field than those on non-aromatic sp^2 carbons. This is an important way of detecting aromaticity. By the same mechanism, the signals of protons located near the ring axis are shifted upfield.

Aromatic molecules are able to interact with each other in so-called π–π stacking: The π systems form two parallel rings overlap in a face-to-face orientation. Aromatic molecules are also able to interact with each other in an edge-to-face orientation: The slight positive charge of the substituents on the ring atoms of one molecule are attracted to the slight negative charge of the aromatic system on another molecule.

Planar monocyclic molecules containing $4n$ π-electrons are called antiaromatic and are, in general, unstable. Molecules that could be antiaromatic will tend to change from this electronic or conformation, thereby becoming non-aromatic. For example, cyclooctatetraene (COT) distorts out of planarity, breaking π overlap between adjacent double bonds. Recent studies have determined that cyclobutadiene adopts an asymmetric, rectangular configuration in which single and double bonds indeed alternate, with no resonance; the single bonds are markedly longer than the double bonds, reducing unfavorable p-orbital overlap. This reduction of symmetry lifts the degeneracy of the two formerly non-bonding molecular orbitals, which by Hund›s rule forces the two unpaired electrons into a new, weakly bonding orbital (and also creates a weakly antibonding orbital). Hence, cyclobutadiene is non-aromatic; the strain of the asymmetric configuration outweighs the anti-aromatic destabilization that would afflict the symmetric, square configuration.

Importance of Aromatic Compounds

Aromatic compounds play key roles in the biochemistry of all living things. The four aromatic amino acids histidine, phenylalanine, tryptophan, and tyrosine each serve as one of the 20 basic building-blocks of proteins. Further, all 5 nucleotides (adenine, thymine, cytosine, guanine, and uracil) that make up the sequence of the genetic code in DNA and RNA are aromatic purines or pyrimidines. The molecule heme contains an aromatic system with 22 π-electrons. Chlorophyll also has a similar aromatic system.

Aromatic compounds are important in industry. Key aromatic hydrocarbons of commercial interest are benzene, toluene, *ortho*-xylene and *para*-xylene. About 35 million tonnes are produced worldwide every year. They are extracted from complex mixtures obtained by the refining of oil or by distillation of coal tar, and are used to produce a range of important chemicals and polymers, including styrene, phenol, aniline, polyester and nylon.

Types of Aromatic Compounds

The overwhelming majority of aromatic compounds are compounds of carbon, but they need not be hydrocarbons.

Neutral Homocyclics

Benzene, as well as most other annulenes (cyclodecapentaene excepted) with the formula C_nH_n where n is an even number, such as cyclotetradecaheptaene.

Heterocyclics

In heterocyclic aromatics (heteroaromatics), one or more of the atoms in the aromatic ring is of an element other than carbon. This can lessen the ring›s aromaticity, and thus (as in the case of furan) increase its reactivity. Other examples include pyridine, pyrazine, imidazole, pyrazole, oxazole, thiophene, and their benzannulated analogs (benzimidazole, for example). In all these examples, the number of π-electrons is 6, due to the π-electrons from the double bonds as well as the two electrons from any lone pair that is in the p-orbital that is in the plane of the aromatic π system. For example, in pyridine, the five sp^2-hybridized carbons each have a p-orbital that is perpendicular to the plane of the ring, and each of these p-orbitals contains one π-electron. Additionally, the nitrogen atom is also sp^2-hybridized and has one electron in a p-orbital, which adds up to 6 p-electrons, thus making pyridine aromatic. The lone pair on the nitrogen is not part of the aromatic π system. Pyrrole and imidazole are both five membered aromatic rings that contain heteroatoms. In pyrrole, each of the four sp^2-hybridized carbons contributes one π-electron, and the nitrogen atom is also sp^2-hybridized and contributes two π-electrons from its lone pair, which occupies a p-orbital. In imidazole, both nitrogens are sp^2-hybridized; the one in the double bond contributes one electron and the one which is not in the double bond and is in a lone pair contributes two electrons to the π system.

Fused Aromatics and Polycyclics

Polycyclic aromatic hydrocarbons are molecules containing two or more simple aromatic rings fused together by sharing two neighboring carbon atoms (see also simple aromatic rings). Examples are naphthalene, anthracene, and phenanthrene. In fused aromatics, not all carbon–carbon bonds are necessarily equivalent, as the electrons are not delocalized over the entire molecule. The aromaticity of these molecules can be explained using their orbital picture. Like benzene and other monocyclic aromatic molecules, polycyclics have a cyclic conjugated pi system with p-orbital overlap above and below the plane of the ring.

Substituted Aromatics

Many chemical compounds are aromatic rings with other functional groups attached. Examples include trinitrotoluene (TNT), acetylsalicylic acid (aspirin), paracetamol, and the nucleotides of DNA.

Aromatic Ions

Aromatic molecules need not be neutral molecules. Ions that satisfy Huckel's rule of $4n + 2$ π-electrons in a planar, cyclic, conjugated molecule are considered to be aromatic ions. For example, the cyclopentadienyl anion and the cycloheptatrienylium cation are both considered to be aromatic ions, and the azulene molecule can be approximated as a combination of both.

In order to convert the atom from sp^3 to sp^2, a carbocation, carbanion, or carbon radical must be formed. These leave sp^2-hybridized carbons that can partake in the π system of an aromatic molecule. Like neutral aromatic compounds, these compounds are stable and form easily. The cyclopentadienyl anion is formed very easily and thus 1,3-cyclopentadiene is a very acidic hydrocarbon with a pK_a of 16. Other examples of aromatic ions include the cyclopropenium cation (2 π-electrons) and cyclooctatetraenyl dianion (10 pi electrons).

Atypical Aromatic Compounds

Aromaticity also occurs in compounds that are not carbocyclic or heterocyclic; inorganic six-membered-ring compounds analogous to benzene have been synthesized. For example, borazine is a six-membered ring composed of alternating boron and nitrogen atoms, each with one hydrogen attached. It has a delocalized π system and undergoes electrophilic substitution reactions appropriate to aromatic rings rather than reactions expected of non-aromatic molecules.

Quite recently, the aromaticity of planar Si6−5 rings occurring in the Zintl phase $Li_{12}Si_7$ was experimentally evinced by Li solid-state NMR Metal aromaticity is believed to exist in certain clusters of aluminium, for example.

Homoaromaticity is the term used to describe systems where conjugation is interrupted by a single sp^3 hybridized carbon atom.

Möbius aromaticity occurs when a cyclic system of molecular orbitals, formed from p_π atomic orbitals and populated in a closed shell by $4n$ (n is an integer) electrons, is given a single half-twist to form a Möbius strip. A π system with $4n$ electrons in a flat (non-twisted) ring would be antiaromatic, and therefore highly unstable, due to the symmetry of the combinations of p atomic orbitals. By twisting the ring, the symmetry of the system changes and becomes allowed (see also Möbius–Hückel concept for details). Because the twist can be left-handed or right-handed, the resulting Möbius aromatics are *dissymmetric* or chiral. But as of 2012, no Möbius aromatic molecules had been synthesized. Aromatics with two half-twists corresponding to the paradromic topologies were first suggested by Johann Listing. In carbo-benzene the ring bonds are extended with alkyne and allene groups.

Y-aromaticity is a concept developed to explain the extraordinary stability and high basicity of the guanidinium cation. Guanidinium is not a ring molecule, and is cross-conjugated rather than a linear π system, but is reported to have its six π-electrons delocalized over the whole molecule. The concept is controversial and some authors emphasize different effects.

POLYCYCLIC AROMATIC HYDROCARBON

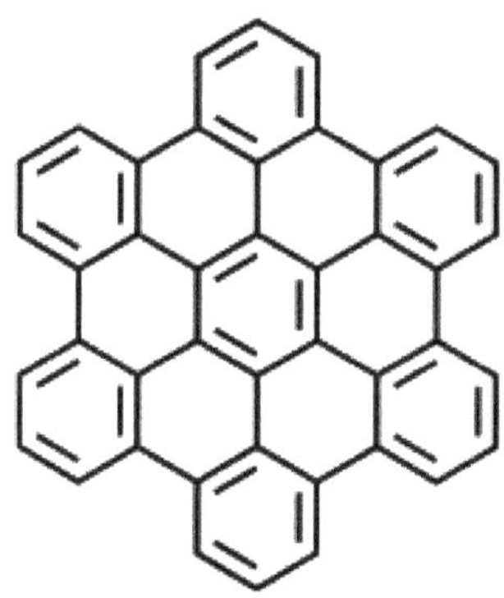

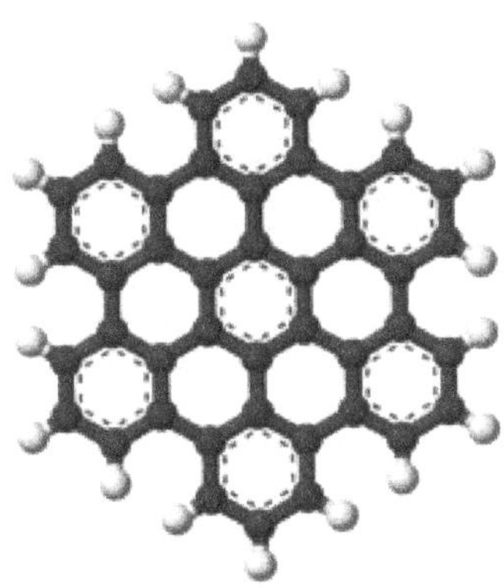

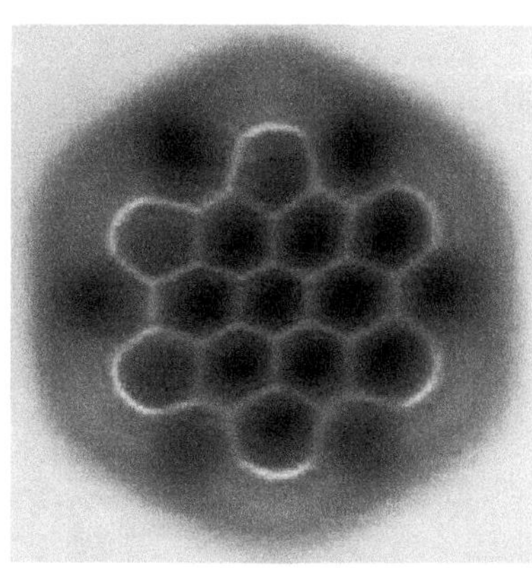

Three representations of hexabenzocoronene, a polycylic aromatic hydrocarbon. Top: standard line-angle schematic, where carbon atoms are represented by the vertices of the hexagons and hydrogen atoms are inferred. Middle: ball-and-stick model showing all carbon and hydrogen atoms. Bottom: atomic force microscopy image.

Polycyclic aromatic hydrocarbons (**PAHs**, also ***polyaromatic hydrocarbons***) are hydrocarbons—organic compounds containing only carbon and hydrogen—that are composed of multiple aromatic rings (organic rings in which the electrons are delocalized). Formally, the class is further defined as lacking further branching substituents on these ring structures. **Polynuclear aromatic hydrocarbons** (**PNAs**) are a subset of PAHs that have fused aromatic rings, that is, rings that share one or more sides. The simplest such chemicals are naphthalene, having two aromatic rings, and the three-ring compounds anthracene and phenanthrene.

PAHs are neutral, nonpolar molecules found in coal and in tar deposits. They are produced as well by incomplete combustion of organic matter (e.g., in engines and incinerators, when biomass burns in forest fires, etc.).

PAHs may also be abundant in the universe, and are conjectured to have formed as early as the first couple of billion years after the Big Bang, in association with formation of new stars and exoplanets. Some studies suggest that PAHs account for a significant percentage of all carbon in the universe, and PAHs are discussed as possible starting materials for abiologic syntheses of materials required by the earliest forms of life.

Nomenclature, Structure, Properties

Nomenclature and Structure

The tricyclic species phenanthrene and anthracene represent the starting members of the PAHs. Smaller molecules, such as benzene, are not PAHs, and PAHs are not generally considered to contain heteroatoms or carry substituents.

PAHs with five or six-membered rings are most common. Those composed only of six-membered rings are called *alternant* PAHs, which include *benzenoid* PAHs. The following are examples of PAHs that vary in the number and arrangement of their rings:

Examples of PAH compounds

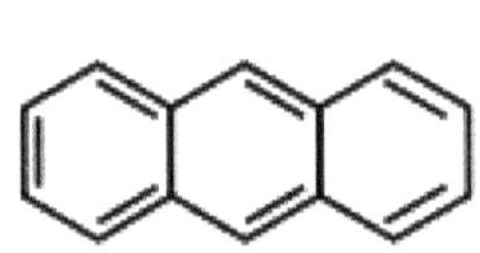

Anthracene

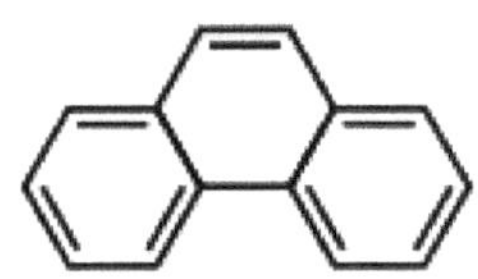

Phenanthrene

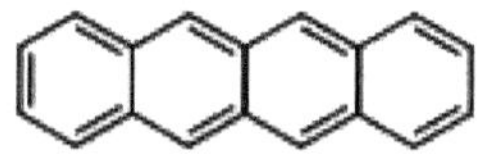

Tetracene

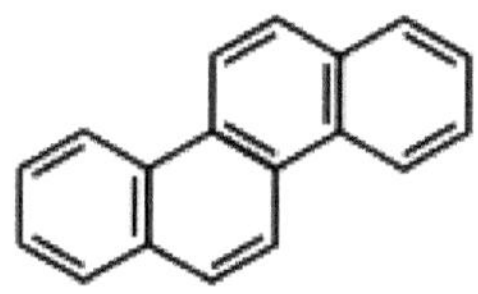

Chrysene

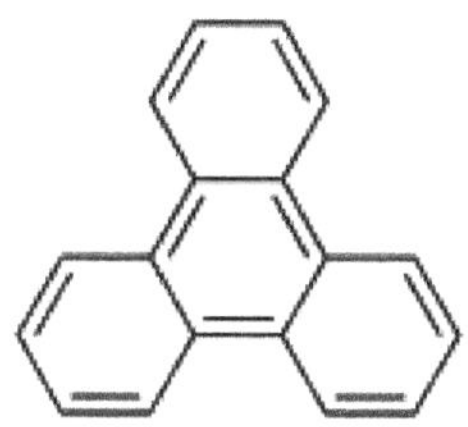

Triphenylene

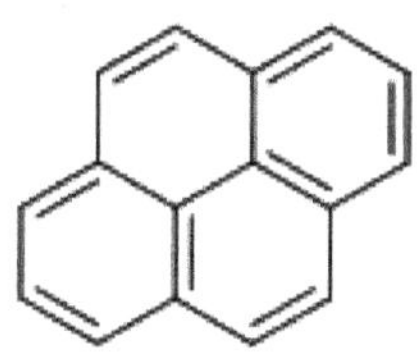

Pyrene

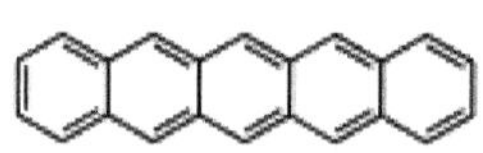

Pentacene

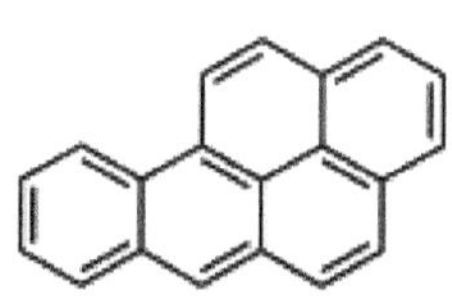

Benzo[*a*]pyrene

Corannulene

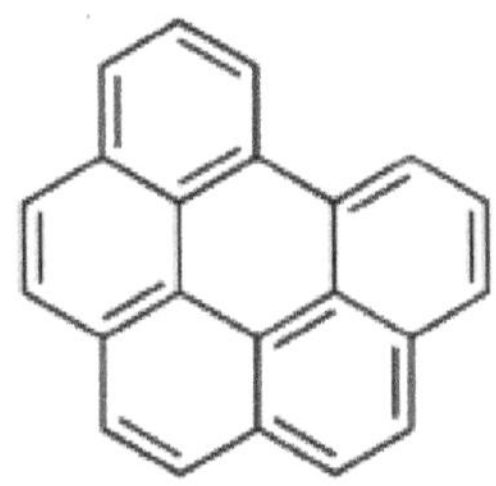

Benzo[*ghi*]perylene

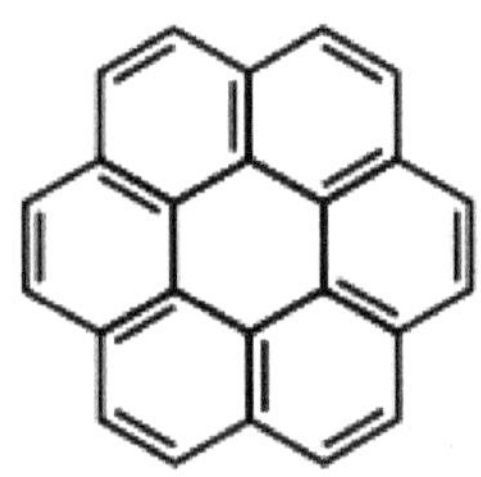

Coronene

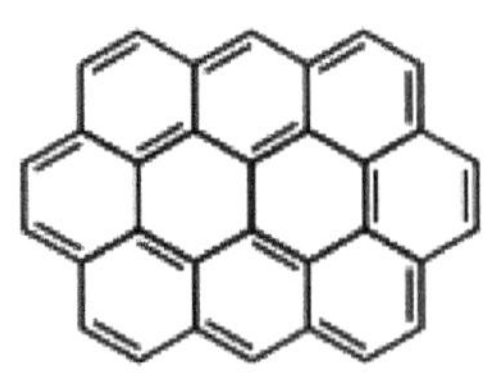

Ovalene

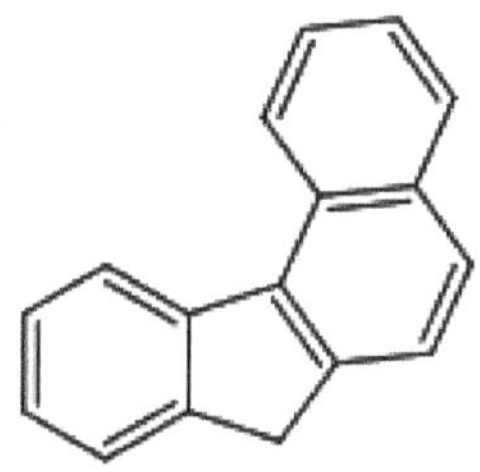

Benzo[*c*]fluorene

Physicochemical Properties and Bonding

PAHs are nonpolar and lipophilic. PAHs are insoluble in water. The larger members are also poorly soluble in organic solvents as well as lipids. They are usually colorless.

Although PAHs clearly are aromatic compounds, the degree of aromaticity can be different for each ring segment. According to *Clar›s rule* (formulated by Erich Clar in 1964) for PAHs the resonance structure with the largest number of disjoint aromatic п-sextets—i.e. benzene-like moieties—is the most important for the characterization of the properties.

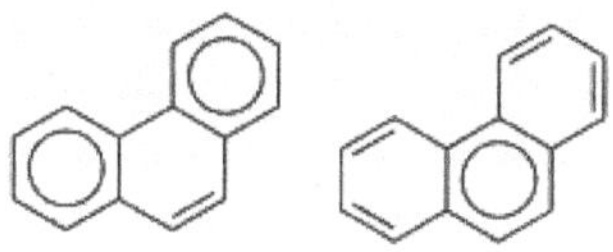	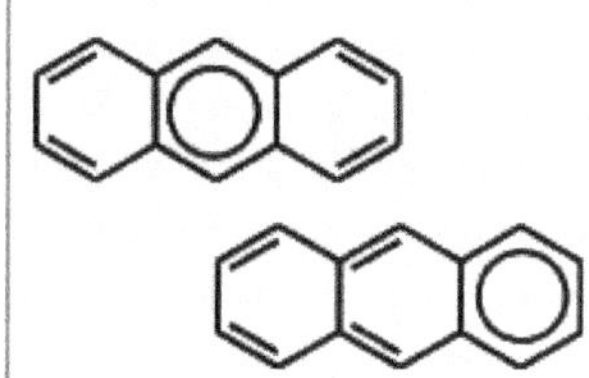	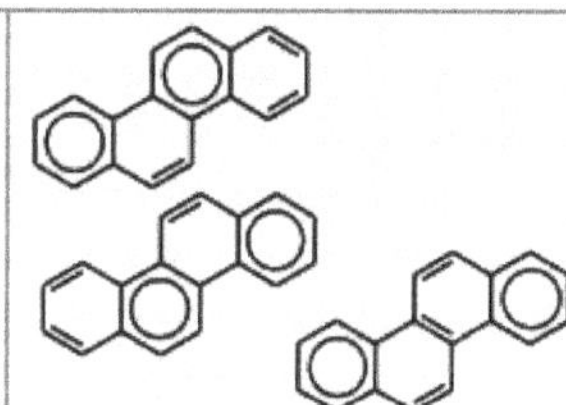
Phenanthrene	**Anthracene**	**Chrysene**

For example, in phenanthrene one Clar structure has two sextets at the extremities, while the other resonance structure has just one central sextet; therefore in this molecule the outer rings have greater aromatic character whereas the central ring is less aromatic and therefore more reactive. In contrast, in anthracene the resonance structures have one sextet, which can be at any of the three rings, and the aromaticity spreads out more evenly across the whole molecule. This difference in number of sextets is reflected in the UV absorbance spectra of these two isomers; phenanthrene has a highest wavelength absorbance around 290 nm, while anthracene has highest wavelength bands around 380 nm. Three Clar structures with two sextets each are present in chrysene. Superposition of these structures reveals that the aromaticity in the outer rings is greater (each has a sextet in two of the three Clar structures) compared to the inner rings (each has a sextet in only one of the three).

Sources and Distribution

Polycyclic aromatic hydrocarbons are primarily found in natural sources such as creosote. They can result from the incomplete combustion of organic matter. PAHs can also be produced geologically when organic sediments are chemically transformed into fossil fuels such as oil and coal. The dominant sources of PAHs in the environment are thus from human activity: Wood-burning and combustion of other biofuels such as dung or crop residues contribute more than half of annual global PAH emissions, particularly due to biofuel use in India and China. As of 2004, industrial processes and the extraction and use of fossil fuels made up slightly more than one quarter of global PAH emissions, dominating outputs in industrial countries such as the United States. Wild fires are another notable source. Substantially higher outdoor air, soil, and water concentrations of PAHs have been measured in Asia, Africa, and Latin America than in Europe, Australia, and the U.S./Canada.

PAHs are typically found as complex mixtures. Lower-temperature combustion, such as tobacco smoking or wood-burning, tends to generate low molecular weight PAHs, whereas high-temperature industrial processes typically generate PAHs with higher molecular weights.

In the Aqueous Environment

Most PAHs are insoluble in water, which limits their mobility in the environment. Aqueous solubility of PAHs decreases approximately logarithmically as molecular mass increases.

Two-ring PAHs, and to a lesser extent three-ring PAHs, dissolve in water, making them more available for biological uptake and degradation. Further, two- to four-ring PAHs volatilize sufficiently to appear in the atmosphere predominantly in gaseous form, although the physical state of four-ring PAHs can depend on temperature. In contrast, compounds with five or more rings have low solubility in water and low volatility; they are therefore predominantly in solid state, bound to particulate air pollution, soils, or sediments. In solid state, these compounds are less accessible for biological uptake or degradation, increasing their persistence in the environment.

Human Exposure

Human exposure varies across the globe and depends on factors such as smoking rates, fuel types in cooking, and pollution controls on power plants, industrial processes, and vehicles. Developed countries with stricter air and water pollution controls, cleaner sources of cooking (i.e., gas and electricity vs. coal or biofuels), and prohibitions of public smoking tend to have lower levels of PAH exposure, while developing and undeveloped countries tend to have higher levels.

Burning solid fuels such as coal and biofuels in the home for cooking and heating is a dominant global source of PAH emissions that in developing countries leads to high levels of exposure to indoor particulate air pollution containing PAHs, particularly for women and children who spend more time in the home or cooking.

In industrial countries, people who smoke tobacco products, or who are exposed to second-hand smoke, are among the most highly exposed groups; tobacco smoke contributes to 90% of indoor PAH levels in the homes of smokers. For the general population in developed countries, the diet is otherwise the dominant source of PAH exposure, particularly from smoking or grilling meat or consuming PAHs deposited on plant foods, especially broad-leafed vegetables, during growth. PAHs are typically at low concentrations in drinking water.

A wood-burning open-air cook stove. Smoke from solid fuels like wood is a large source of PAHs globally.

Smog in Cairo, Egypt. Particulate air pollution, including smog, is a substantial avenue for human exposure to PAHs.

Emissions from vehicles such as cars and trucks can be a substantial outdoor source of PAHs in particulate air pollution.Geographically, major roadways are thus sources of PAHs, which may distribute in the atmosphere or deposit nearby. Catalytic converters are estimated to reduce PAH emissions from gasoline-fired vehicles by 25-fold.

People can also be occupationally exposed during work that involves fossil fuels or their derivatives, wood burning, carbon electrodes, or exposure to diesel exhaust. Industrial activity that can produce and distribute PAHs includes aluminum, iron, and steel manufacturing; coal gasification, tar distillation, shale oil extraction; production of coke, creosote, carbon black, and calcium carbide; road paving and asphalt manufacturing; rubber tire production; manufacturing or use of metal working fluids; and activity of coal or natural gas power stations.

Environmental Distribution and Degradation

Oil on a beach after a 2007 oil spill in Korea.

PAHs typically disperse from urban and suburban non-point sources through road run-off, sewage, and atmospheric circulation and subsequent deposition of particulate air pollution. Soil and river sediment near industrial sites such as creosote manufacturing facilities can be highly contaminated with PAHs. Oil spills, creosote, coal mining dust, and other fossil fuel sources can also distribute PAHs in the environment.

Two- and three-ring PAHs can disperse widely while dissolved in water or as gases in the atmosphere, while PAHs with higher molecular weights can disperse locally or regionally adhered to particulate matter that is suspended in air or water until the particles land or settle out of the water column. PAHs have a strong affinity for organic carbon, and thus highly organic sediments in rivers, lakes, and the ocean can be a substantial sink for PAHs.

Algae and some invertebrates such as protozoans, mollusks, and many polychaetes have limited ability to metabolize PAHs and bioaccumulate disproportionate concentrations of PAHs in their tissues; however, PAH metabolism can vary substantially across invertebrate species. Most vertebrates metabolize and excrete PAHs relatively rapidly. Tissue concentrations of PAHs do not increase (biomagnify) from the lowest to highest levels of food chains.

PAHs transform slowly to a wide range of degradation products. Biological degradation by microbes is a dominant form of PAH transformation in the environment. Soil-consuming invertebrates such as earthworms speed PAH degradation, either through direct metabolism or by improving the conditions for microbial transformations. Abiotic degradation in the atmosphere and the top layers of surface waters can produce nitrogenated, halogenated, hydroxylated, and oxygenated PAHs; some of these compounds can be more toxic, water-soluble, and mobile than their parent PAHs.

Minor Sources

Volcanic eruptions may emit PAHs. Certain PAHs such as perylene can also be generated in anaerobic sediments from existing organic material, although it remains undetermined whether abiotic or microbial processes drive their production.

Human Health

Cancer is a primary human health risk of exposure to PAHs. Exposure to PAHs has also been linked with cardiovascular disease and poor fetal development.

Cancer

PAHs have been linked to skin, lung, bladder, liver, and stomach cancers in well-established animal model studies. Specific compounds classified by various agencies as possible or probable human carcinogens are identified in the section "Regulation and Oversight below.

Historical Significance

An eighteenth century drawing of chimney sweeps.

Historically, PAHs contributed substantially to our understanding of adverse health effects from exposures to environmental contaminants, including chemical carcinogenesis. In 1775, Percivall Pott, a surgeon at St. Bartholomew's Hospital in London, observed that scrotal cancer was unusually common in chimney sweepers and proposed the cause as occupational exposure to soot. A century later, Richard von Volkmann reported increased skin cancers in workers of the coal tar industry of Germany, and by the early 1900s increased rates of cancer from exposure to soot and coal tar was widely accepted. In 1915, Yamigawa and Ichicawa were the first to experimentally produce cancers, specifically of the skin, by topically applying coal tar to rabbit ears.

In 1922, Ernest Kennaway determined that the carcinogenic component of coal tar mixtures was an organic compound consisting of only C and H. This component was later linked to a characteristic fluorescent pattern that was similar but not identical to benz[*a*]anthracene, a PAH that was subsequently demonstrated to cause tumors. Cook, Hewett and Hieger then linked the specific spectroscopic fluorescent profile of benzo[*a*] pyrene to that of the carcinogenic component of coal tar, the first time that a specific compound from an environmental mixture (coal tar) was demonstrated to be carcinogenic.

In the 1930s and later, epidemiologists from Japan, England, and the U.S., including Richard Doll and various others, reported greater rates of death from lung cancer following occupational exposure to PAH-rich environments among workers in coke ovens and coal carbonization and gasification processes.

Mechanisms of Carcinogenesis

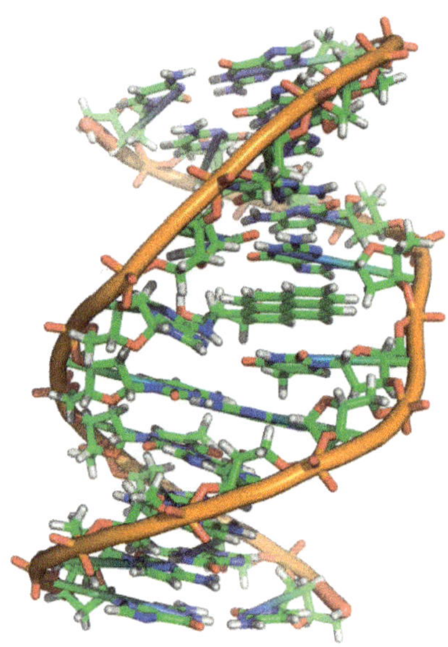

An adduct formed between a DNA strand and an epoxide derived from a benzo[*a*]pyrene molecule (center); such adducts may interfere with normal DNA replication.

The structure of a PAH influences whether and how the individual compound is carcinogenic. Some carcinogenic PAHs are genotoxic and induce mutations that initiate cancer; others are not genotoxic and instead affect cancer promotion or progression.

PAHs that affect cancer initiation are typically first chemically modified by enzymes into metabolites that react with DNA, leading to mutations. When the DNA sequence is altered in genes that regulate cell replication, cancer can result. Mutagenic PAHs, such as benzo[*a*]pyrene, usually have four or more aromatic rings as well as a "bay region", a structural pocket that increases reactivity of the molecule to the metabolizing enzymes. Mutagenic metabolites of PAHs include diol epoxides, quinones, and radical PAH cations. These metabolites can bind to DNA at specific sites, forming bulky complexes called DNA adducts that can be stable or unstable. Stable adducts may lead to DNA replication errors, while unstable adducts react with the DNA strand, removing a purine base (either adenine or guanine). Such mutations, if they are not repaired, can transform genes encoding for normal cell signaling proteins into cancer-causing oncogenes. Quinones can also repeatedly generate reactive oxygen species that may independently damage DNA.

Enzymes in the cytochrome family (CYP1A1, CYP1A2, CYP1B1) metabolize PAHs to diol epoxides. PAH exposure can increase production of the cytochrome enzymes, allowing the enzymes to convert PAHs into mutagenic diol epoxides at greater rates. In this pathway, PAH molecules bind to the aryl hydrocarbon receptor (AhR) and activate it as a transcription factor that increases production of the cytochrome enzymes. The activity of these enzymes may at times conversely protect against PAH toxicity, which is not yet well understood.

Low molecular weight PAHs, with 2 to 4 aromatic hydrocarbon rings, are more potent as co-carcinogens during the promotional stage of cancer. In this stage, an initiated cell (i.e., a cell that has retained a carcinogenic mutation in a key gene related to cell replication) is removed from growth-suppressing signals from its neighboring cells and begins to clonally replicate. Low molecular weight PAHs that have bay or bay-like regions can dysregulate gap junction channels, interfering with intercellular communication, and also affect mitogen-activated protein kinases that activate transcription factors involved in cell proliferation. Closure of gap junction protein channels is a normal precursor to cell division. Excessive closure of these channels after exposure to PAHs results in removing a cell from the normal growth-regulating signals imposed by its local community of cells, thus allowing initiated cancerous cells to replicate. These PAHs do not need to be enzymatically metabolized first. Low molecular weight PAHs are prevalent in the environment, thus posing a significant risk to human health at the promotional phases of cancer.

Cardiovascular Disease

Adult exposure to PAHs has been linked to cardiovascular disease. PAHs are among the complex suite of contaminants in cigarette smoke and particulate air pollution and may contribute to cardiovascular disease resulting from such exposures.

In laboratory experiments, animals exposed to certain PAHs have shown increased development of plaques (atherogenesis) within arteries. Potential mechanisms for the pathogenesis and development of atherosclerotic plaques may be similar to the mechanisms involved in the carcinogenic and mutagenic properties of PAHs. A leading hypothesis is that PAHs may activate the cytochrome enzyme CYP1B1 in vascular smooth muscle cells. This enzyme then metabolically processes the PAHs to quinone metabolites that bind to DNA in reactive adducts that remove purine bases. The resulting mutations may contribute to unregulated growth of vascular smooth muscle cells or to their migration to the inside of the artery, which are steps in plaque formation. These quinone metabolites also generate reactive oxygen species that may alter the activity of genes that affect plaque formation.

Oxidative stress following PAH exposure could also result in cardiovascular disease by causing inflammation, which has been recognized as an important factor in the development of atherosclerosis and cardiovascular disease. Biomarkers of exposure to PAHs in humans have been associated with inflammatory biomarkers that are recognized as important predictors of cardiovascular disease, suggesting that oxidative stress resulting from exposure to PAHs may be a mechanism of cardiovascular disease in humans.

Developmental Impacts

Multiple epidemiological studies of people living in Europe, the United States, and China have linked in utero exposure to PAHs, through air pollution or parental occupational exposure, with poor fetal growth, reduced immune function, and poorer neurological development, including lower IQ.

Regulation and Oversight

Some governmental bodies, including the European Union as well as NIOSH and the Environmental Protection Agency in the US, regulate concentrations of PAHs in air, water, and soil. The European Commission has restricted concentrations of 8 carcinogenic PAHs in consumer products that contact the skin or mouth.

Priority polycyclic aromatic hydrocarbons identified by the US EPA, the US Agency for Toxic Substances and Disease Registry (ATSDR), and the European Food Safety Authority (EFSA) due to their carcinogenicity or genotoxicity and/or ability to be monitored are the following:

Compound	Agency
acenaphthene	EPA, ATSDR
acenaphthylene	EPA, ATSDR
anthracene	EPA, ATSDR
benz[*a*]anthracene[A]	EPA, ATSDR, EFSA
benzo[*b*]fluoranthene[A]	EPA, ATSDR, EFSA
benzo[*j*]fluoranthene	ATSDR, EFSA
benzo[*k*]fluoranthene[A]	EPA, ATSDR, EFSA
benzo[*c*]fluorene	EFSA
benzo[*g,h,i*]perylene[A]	EPA, ATSDR, EFSA

benzo[*a*]pyrene[A]	EPA, ATSDR, EFSA
benzo[*e*]pyrene	ATSDR
chrysene[A]	EPA, ATSDR, EFSA
coronene	ATSDR

Compound	**Agency**
cyclopenta[*c,d*]pyrene	EFSA
dibenz[*a,h*]anthracene[A]	EPA, ATSDR, EFSA
dibenzo[*a,e*]pyrene	EFSA
dibenzo[*a,h*]pyrene	EFSA
dibenzo[*a,i*]pyrene	EFSA
dibenzo[*a,l*]pyrene	EFSA
fluoranthene	EPA, ATSDR
fluorene	EPA, ATSDR
indeno[1,2,3-*c,d*]pyrene[A]	EPA, ATSDR, EFSA
5-methylchrysene	EFSA
naphthalene	EPA
phenanthrene	EPA, ATSDR
pyrene	EPA, ATSDR

[A] Considered probable or possible human carcinogens by the US EPA, the European Union, and/or the International Agency for Research on Cancer (IARC).

Detection and Optical Properties

A spectral database exists for tracking polycyclic aromatic hydrocarbons (PAHs) in the universe.greatly upgraded database Detection of PAHs in materials is often done using gas chromatography-mass spectrometry or liquid chromatography with ultraviolet-visible or fluorescence spectroscopic methods or by using rapid test PAH indicator strips.

PAHs possess very characteristic UV absorbance spectra. These often possess many absorbance bands and are unique for each ring structure. Thus, for a set of isomers, each isomer has a different UV absorbance spectrum than the others. This is particularly useful in the identification of PAHs. Most PAHs are also fluorescent, emitting characteristic wavelengths of light when they are excited (when the molecules absorb light). The extended pi-electron electronic structures of PAHs lead to these spectra, as well as to certain large PAHs also exhibiting semi-conducting and other behaviors.

Origins of Life

The Spitzer Space Telescope captures PAH spectral lines, producing this image of nebulosity in a stellar nursery.

PAHs may be abundant in the universe. They seem to have been formed as early as a couple of billion years after the Big Bang, and are associated with new stars and exoplanets. More than 20% of the carbon in the universe may be associated with PAHs. PAHs are considered possible starting material for the earliest forms of life. Light emitted by the Red Rectangle nebula and found spectral signatures that suggest the presence of anthracene and pyrene. This report was considered a controversial hypothesis that as nebulae of the same type as the Red Rectangle approach the ends of their lives, convection currents cause carbon and hydrogen in the nebulae›s core to get caught in stellar winds, and radiate outward. As they cool, the atoms supposedly bond to each other in various ways and eventually form particles of a million or more atoms. Witt and his team inferred that PAHs —which may have been vital in the formation of early life on Earth— in a nebula, by necessity they must originate in nebulae.

Two extremely bright stars illuminate a mist of PAHs in this Spitzer image.

More recently, fullerenes (or "buckyballs), have been detected in other nebulae. Fullerenes are also implicated in the origin of life; according to astronomer Letizia

Stanghellini, It›s possible that buckyballs from outer space provided seeds for life on Earth. In September 2012, NASA scientists reported results of analog studies *in vitro* that PAHs, subjected to interstellar medium (ISM) conditions, are transformed, through hydrogenation, oxygenation, and hydroxylation, to more complex organics—a step along the path toward amino acids and nucleotides, the raw materials of proteins and DNA, respectively. Further, as a result of these transformations, the PAHs lose their spectroscopic signature which could be one of the reasons "for the lack of PAH detection in interstellar ice grains, particularly the outer regions of cold, dense clouds or the upper molecular layers of protoplanetary disks.

PAHs have been detected in the upper atmosphere of Titan, the largest moon of the planet Saturn.

REFERENCES

1. G.P. Moss IUPAC nomenclature for fused-ring systems
2. Fetzer, J. C. (2000). The Chemistry and Analysis of the Large Polycyclic Aromatic Hydrocarbons. Polycyclic Aromatic Compounds. New York: Wiley. **27** *(2): 143–162.* doi:10.1080/10406630701268255. ISBN 0-471-36354-5.
3. Harvey, R.G. (1998). Environmental Chemistry of PAHs. PAHs and Related Compounds: Chemistry. The Handbook of Environmental Chemistry. Springer. pp. 1–54. ISBN 9783540496977.
4. Xinliang Feng, Wojciech Pisula, and Klaus Müllen Large polycyclic aromatic hydrocarbons: Synthesis and discotic organizationPure Appl. Chem.,2009, Vol. 81, No. 12, pp. 2203–2224.doi:10.1351/PAC-CON-09-07-07
5. Clar, E. (1964). Polycyclic Hydrocarbons. New York: Academic Press. LCCN 63012392.
6. Portella, G.; Poater, J.; Solà, M. (2005). Assessment of Clar›s aromatic π-sextet rule by means of PDI, NICS and HOMA indicators of local aromaticity. Journal of Physical Organic Chemistry. **18** *(8): 785–91. doi:10.1002/poc.938.*
7. Anja Sörensen and Bodo Wichert Asphalt and Bitumen in Ullmann›s Encyclopedia of Industrial Chemistry Wiley-VCH, Weinheim, 2009. doi:10.1002/14356007.a03_169.pub2http://www.qrpoil.com/site/?bitumen
8. Ravindra, K; Sokhi, R; Van Grieken, R (2008). Atmospheric polycyclic aromatic hydrocarbons: Source attribution, emission factors and regulation. Atmospheric Environment. **42** *(13): 2895–2921. doi:10.1016/j.atmosenv.2007.12.010. ISSN 1352-2310. Retrieved 2014-08-12.*
9. Ramesh, A.; Archibong, A.; Hood, D.B.; Guo, Z.; Loganathan, B. G. (2011). Global environmental distribution and human health effects of polycyclic aromatic hydrocarbons. Global Contamination Trends of Persistent Organic Chemicals. Boca Raton, FL: CRC Press. pp. 97–126. ISBN 978-1-4398-3831-0.
10. Tobiszewski, M; Namieśnik, J (2012). PAH diagnostic ratios for the identification of pollution emission sources. Environmental Pollution. **162**: 110–119.

doi:10.1016/j.envpol.2011.10.025. ISSN 0269-7491. PMID 22243855. Retrieved 2015-05-12.

11. Choi, H.; Harrison, R.; Komulainen, H.; Delgado Saborit, J. (2010). "Polycyclic aromatic hydrocarbons". WHO Guidelines for Indoor Air Quality: Selected Pollutants. Geneva: World Health Organization.
12. Johnsen, Anders R.; Wick, Lukas Y.; Harms, Hauke (2005). "Principles of microbial PAH-degradation in soil". Environmental Pollution. **133** *(1): 71–84. doi:10.1016/j.envpol.2004.04.015. ISSN 0269-7491. PMID 15327858. Retrieved 2015-05-07.*
13. Mackay, D.; Callcott, D. (1998). Partitioning and physical chemical properties of PAHs. In A. Neilson (ed.). PAHs and Related Compounds. The Handbook of Environmental Chemistry. Springer Berlin Heidelberg. pp. 325–345. ISBN 978-3-642-08286-3.
14. Atkinson, R.; Arey, J. (1994-10-01). Atmospheric chemistry of gas-phase polycyclic aromatic hydrocarbons: formation of atmospheric mutagens. Environmental Health Perspectives. **102**: 117–126. doi:10.2307/3431940. ISSN 0091-6765. JSTOR 3431940. PMC 1566940. PMID 7821285.
15. Srogi, K. (2007-11-01). Monitoring of environmental exposure to polycyclic aromatic hydrocarbons: a review. Environmental Chemistry Letters. **5** *(4): 169–195.* doi:10.1007/s10311-007-0095-0. ISSN 1610-3661. Retrieved 2014-08-20.
16. Haritash, A. K.; Kaushik, C. P. (2009). Biodegradation aspects of polycyclic aromatic hydrocarbons (PAHs): A review. Journal of Hazardous Materials. **169** *(1–3): 1–15. doi:10.1016/j.jhazmat.2009.03.137. ISSN 0304-3894. PMID 19442441. Retrieved 2014-08-20.*
17. Choi, H.; Harrison, R.; Komulainen, H.; Delgado Saborit, J. (2010). Polycyclic aromatic hydrocarbons. WHO Guidelines for Indoor Air Quality: Selected Pollutants. Geneva: World Health Organization.
18. Kim, K-H; Jahan, S A; Kabir, E (2011). A review of diseases associated with household air pollution due to the use of biomass fuels. Journal of Hazardous Materials. **192** *(2): 425–431. doi:10.1016/j.jhazmat.2011.05.087. ISSN 0304-3894. PMID 21705140. Retrieved 2015-12-11.*
19. Phillips, D H (1999). Polycyclic aromatic hydrocarbons in the diet. Mutation Research/Genetic Toxicology and Environmental Mutagenesis. **443** *(1–2): 139–147.* doi:10.1016/S1383-5742(99)00016-2. ISSN 1383-5718. Retrieved 2015-02-05.
20. Srogi, K. (2007). Monitoring of environmental exposure to polycyclic aromatic hydrocarbons: a review. Environmental Chemistry Letters. **5** *(4): 169–195.* doi:10.1007/s10311-007-0095-0. ISSN 1610-3661. Retrieved 2014-08-20.
21. Boffetta, P; Jourenkova, N; Gustavsson, P (1997). Cancer risk from occupational and environmental exposure to polycyclic aromatic hydrocarbons. Cancer Causes & Control. **8** *(3): 444–472.* doi:10.1023/A:1018465507029. ISSN 1573-7225. Retrieved 2014-12-03.

22. Wagner, M; Bolm-Audorff, U; Hegewald, J; Fishta, A; Schlattmann, P; Schmitt, J; Seidler, A (2015). Occupational polycyclic aromatic hydrocarbon exposure and risk of larynx cancer: a systematic review and meta-analysis. Occupational and Environmental Medicine. **72** *(3): 226–233.* doi:10.1136/oemed-2014-102317. ISSN 1470-7926. PMID 25398415. Retrieved 2015-04-13.

23. Hylland, K (2006). Polycyclic aromatic hydrocarbon (PAH) ecotoxicology in marine ecosystems. Journal of Toxicology and Environmental Health, Part A. **69** *(1–2): 109–123.* doi:10.1080/15287390500259327. ISSN 1528-7394. PMID 16291565. Retrieved 2015-04-02.

24. Achten, C.; Hofmann, T. (2009). Native polycyclic aromatic hydrocarbons (PAH) in coals – A hardly recognized source of environmental contamination. Science of the Total Environment. **407** *(8): 2461–2473. doi:10.1016/j.scitotenv.2008.12.008. ISSN 0048-9697. PMID 19195680. Retrieved 2015-04-09.*

25. Jørgensen, A; Giessing, A M. B.; Rasmussen, L J; Andersen, O (2008). Biotransformation of polycyclic aromatic hydrocarbons in marine polychaetes. Marine Environmental Research. **65** *(2): 171–186. doi:10.1016/j.marenvres.2007.10.001. ISSN 0141-1136. PMID 18023473. Retrieved 2016-01-22.*

26. Johnsen, A. R.; Wick, L. Y.; Harms, H. (2005). Principles of microbial PAH-degradation in soil. Environmental Pollution. **133** *(1): 71–84. doi:10.1016/j.envpol.2004.04.015. ISSN 0269-7491. PMID 15327858. Retrieved 2015-05-07.*

27. Lundstedt, S.; White, P. A.; Lemieux, C. L.; Lynes, K. D.; Lambert, I. B.; *Öberg*, L.; Haglund, P.; Tysklind, M. (2007). Sources, fate, and toxic hazards of oxygenated polycyclic aromatic hydrocarbons (PAHs) at PAH- contaminated sites. AMBIO: A Journal of the Human Environment. **36** *(6): 475–485.* doi:10.1579/0044-7447(2007)36[475:SFATHO]2.0.CO;2. ISSN 0044-7447. Retrieved 2015-04-09.

28. Fu, P. P.; Xia, Q; Sun, X; Yu, H (2012). Phototoxicity and Environmental Transformation of Polycyclic Aromatic Hydrocarbons (PAHs)—Light-Induced Reactive Oxygen Species, Lipid Peroxidation, and DNA Damage. Journal of Environmental Science and Health, Part C. **30** *(1): 1–41. doi:10.1080/10590501.2012.653887. ISSN 1059-0501. PMID 22458855. Retrieved 2015-12-11.*

29. Meyers, P. A.; Ishiwatari, R. (1993). Lacustrine organic geochemistry—an overview of indicators of organic matter sources and diagenesis in lake sediments. Organic Geochemistry. **20** *(7): 867–900.* doi:10.1016/0146-6380(93)90100-P. ISSN 0146-6380. Retrieved 2015-02-04.

30. Silliman, J E; Meyers, P A; Eadie, B J; Val Klump, J (2001). A hypothesis for the origin of perylene based on its low abundance in sediments of Green Bay, Wisconsin. Chemical Geology. **177** *(3–4): 309–322.* doi:10.1016/S0009-2541(00)00415-0. ISSN 0009-2541. Retrieved 2015-02-04.

31. Wakeham, S. G.; Schaffner, C.; Giger, W. (1980). Poly cyclic aromatic hydrocarbons in Recent lake sediments—II. Compounds derived from biogenic precursors during early diagenesis. Geochimica et Cosmochimica Acta. **44** *(3): 415–429.* doi:10.1016/0016-7037(80)90041-1. ISSN 0016-7037. Retrieved 2015-02-04.

32. Bostrom, C.-E.; Gerde, P.; Hanberg, A.; Jernstrom, B.; Johansson, C.; Kyrklund, T.; Rannug, A.; Tornqvist, M.; Victorin, K.; Westerholm, R. (2002). Cancer risk assessment, indicators, and guidelines for polycyclic aromatic hydrocarbons in the ambient air.. *Environmental Health Perspectives.* **110** *(*Suppl 3): 451–488. doi:10.1289/ehp.02110s3451. ISSN 0091-6765. PMC 1241197. PMID 12060843.
33. Loeb, L. A.; Harris, C. C. (2008). Advances in Chemical Carcinogenesis: A Historical Review and Prospective. Cancer Research. **68** *(17): 6863–6872. doi:10.1158/0008-5472.CAN-08-2852. ISSN 0008-5472. PMC 2583449. PMID 18757397.*
34. Dipple, A. (1985). Polycyclic Aromatic Hydrocarbon Carcinogenesis. Polycyclic Hydrocarbons and Carcinogenesis. ACS Symposium Series. **283**. American Chemical Society. pp. 1–17. ISBN 0-8412-0924-3.
35. International Agency for Research on Cancer (1984). Polynuclear Aromatic Compounds, Part 3, Industrial Exposures in Aluminium Production, Coal Gasification, Coke Production, and Iron and Steel Founding (Report). IARC Monographs on the Evaluation of Carcinogenic Risks to Humans. Lyon, France: World Health Organization. pp. 89–92,118–124. Retrieved 2016-02-13.
36. Baird, W. M.; Hooven, L. A.; Mahadevan, B. (2005). Carcinogenic polycyclic aromatic hydrocarbon-DNA adducts and mechanism of action. Environmental and Molecular Mutagenesis. **45** *(2–3): 106–114. doi:10.1002/em.20095. ISSN 1098-2280. PMID 15688365. Retrieved 2014.*Check date values in: |access-date= (help)
37. Slaga, T.J. (1984). Chapter 7: Multistage skin carcinogenesis: A useful model for the study of the chemoprevention of cancer. Acta Pharmacologica et Toxicologica. **55** *(S2): 107–124. doi:10.1111/j.1600-0773.1984.tb02485.x. ISSN 1600-0773. Retrieved 2015-05-06.*
38. Xue, W; Warshawsky, D (2005). Metabolic activation of polycyclic and heterocyclic aromatic hydrocarbons and DNA damage: A review. Toxicology and Applied Pharmacology. **206** *(1): 73–93. doi:10.1016/j.taap.2004.11.006. ISSN 0041-008X. PMID 15963346. Retrieved 2014-08-20.*
39. Shimada, T; Fujii-Kuriyama, Y (2004-01-01). Metabolic activation of polycyclic aromatic hydrocarbons to carcinogens by cytochromes P450 1A1 and 1B1. Cancer Science. **95** *(1): 1–6. doi:10.1111/j.1349-7006.2004.tb03162.x. ISSN 1349-7006. PMID 14720319. Retrieved 2014-10-24.*
40. Androutsopoulos, V.P.; Tsatsakis, A.M.; Spandidos, D.A. (2009). Cytochrome P450 CYP1A1: wider roles in cancer progression and prevention. BMC Cancer. **9** *(1): 187.* doi:10.1186/1471-2407-9-187. ISSN 1471-2407. PMC 2703651. PMID 19531241. Retrieved 2015-05-20.
41. Henkler, F; Stolpmann, K; Luch, Andreas (2012). Exposure to Polycyclic Aromatic Hydrocarbons: Bulky DNA Adducts and Cellular Responses. In A. Luch (ed.). Molecular, Clinical and Environmental Toxicology. Experientia Supplementum. Springer Basel. pp. 107–131. ISBN 978-3-7643-8340-4.

42. Nebert, D. W.; Dalton, T. P.; Okey, A. B.; Gonzalez, F. J. (2004). Role of Aryl Hydrocarbon Receptor-mediated Induction of the CYP1 Enzymes in Environmental Toxicity and Cancer. Journal of Biological Chemistry. **279** *(23): 23847–23850. doi:10.1074/jbc.R400004200. ISSN 1083-351X. PMID 15028720. Retrieved 2014-08-20.*

43. Ramesh, A.; Walker, S. A.; Hood, D. B.; Guillén, M. D.; Schneider, K.; Weyand, E. H. (2004). Bioavailability and risk assessment of orally ingested polycyclic aromatic hydrocarbons. International Journal of Toxicology. **23** *(5): 301–333*. doi:10.1080/10915810490517063. ISSN 1092-874X. PMID 15513831. Retrieved 2015-04-13.

44. Korashy, H. M.; El-Kadi, A. O. S. (2006). The Role of Aryl Hydrocarbon Receptor in the Pathogenesis of Cardiovascular Diseases. Drug Metabolism Reviews. **38** *(3): 411–450.* doi:10.1080/03602530600632063. ISSN 0360-2532. PMID 16877260. Retrieved 2014-08-26.

45. Lewtas, J. (2007). Air pollution combustion emissions: Characterization of causative agents and mechanisms associated with cancer, reproductive, and cardiovascular effects. Mutation Research/Reviews in Mutation Research. The Sources and Potential Hazards of Mutagens in Complex Environmental Matrices - Part II. **636** *(1–3): 95–133. doi:10.1016/j.mrrev.2007.08.003. ISSN 1383-5742. Retrieved 2014-08-12.*

46. Ramos, Kenneth S.; Moorthy, Bhagavatula (2005). Bioactivation of Polycyclic Aromatic Hydrocarbon Carcinogens within the vascular Wall: Implications for Human Atherogenesis. Drug Metabolism Reviews. **37** *(4): 595–610.* doi:10.1080/03602530500251253. ISSN 0360-2532. PMID 16393887. Retrieved 2015-03-27.

47. Kunzli, N.; Tager, I. (2005). Air pollution: from lung to heart (PDF). Swiss Medical Weekly. **135** *(47–48): 697–702. PMID 16511705. Retrieved 2015-12-16.*

48. Ridker, P. M. (2009). C-Reactive Protein: Eighty Years from Discovery to Emergence as a Major Risk Marker for Cardiovascular Disease. Clinical Chemistry. **55** *(2): 209–215. doi:10.1373/clinchem.2008.119214. ISSN 1530-8561. PMID 19095723. Retrieved 2015-12-16.*

49. Rossner Jr, P.; Sram, R.J. (2012). Immunochemical detection of oxidatively damaged DNA. Free Radical Research. **46** *(4): 492–522. doi:10.3109/10715762.2011.632415. ISSN 1071-5762. PMID 22034834. Retrieved 2015-12-17.*

50. Sram, R.J.; Binkova, B.; Dejmek, J.; Bobak, M. (2005). Ambient Air Pollution and Pregnancy Outcomes: A Review of the Literature. Environmental Health Perspectives. **113** *(4): 375–382. doi:10.1289/ehp.6362. ISSN 0091-6765. PMC 1278474. PMID 15811825.*

51. Winans, B; Humble, M; Lawrence, B. P. (2011). Environmental toxicants and the developing immune system: A missing link in the global battle against infectious disease?. *Reproductive Toxicology.* **31** *(3): 327–336. doi:10.1016/j.reprotox.2010.09.004. PMC 3033466. PMID 20851760. Retrieved 2015-04-06.*

52. Wormley, D D; Ramesh, A; Hood, D B (2004). Environmental contaminant–mixture effects on CNS development, plasticity, and behavior. Toxicology and Applied Pharmacology. **197** *(1): 49–65. doi:10.1016/j.taap.2004.01.016. ISSN 0041-008X. PMID 15126074. Retrieved 2014-09-12.*

53. Suades-González, E; Gascon, M; Guxens, M; Sunyer, J (2015). Air Pollution and Neuropsychological Development: A Review of the Latest Evidence. Endocrinology. **156** *(10): 3473–3482. doi:10.1210/en.2015-1403. ISSN 0013-7227. PMC 4588818. PMID 26241071. Retrieved 2016-01-13.*

54. Kim, Ki-Hyun; Jahan, Shamin Ara; Kabir, Ehsanul; Brown, Richard J. C. (2013-10). A review of airborne polycyclic aromatic hydrocarbons (PAHs) and their human health effects. Environment International. **60**: 71–80. doi:10.1016/j.envint.2013.07.019. ISSN 0160-4120. PMID 24013021. Retrieved 2015-04-02. Check date values in: |date= (help)

55. European Union (2013-12-06), Commission Regulation (EU) 1272/2013, retrieved 2016-02-01

56. Keith, Lawrence H. (2014-12-08). The Source of U.S. EPA›s Sixteen PAH Priority Pollutants. Polycyclic Aromatic Compounds. **0** *(2–4): 1–14. doi:10.1080/10406638.2014.892886. ISSN 1040-6638. Retrieved 2015-04-09.*

57. Agency for Toxic Substances and Disease Registry (ATSDR) (1995). Toxicological profile for Polycyclic Aromatic Hydrocarbons (PAHs) (Report). Atlanta, GA: U.S. Department of Health and Human Services, Public Health Service. Retrieved 2015-05-06.

58. ^ Jump up to:[a] [b] EFSA Panel on Contaminants in the Food Chain (CONTAM) (2008). Polycyclic Aromatic Hydrocarbons in Food: Scientific Opinion of the Panel on Contaminants in the Food Chain (Report). Parma, Italy: European Food Safety Authority (EFSA). pp. 1–4.

59. ATSDR, Environmental Medicine; Environmental Health Education (2011-07-01). Toxicity of Polycyclic Aromatic Hydrocarbons (PAHs): Health Effects Associated With PAH Exposure. Retrieved 2016-02-01.

60. Hoover, R. (21 February 2014). Need to Track Organic Nano-Particles Across the Universe? NASA›s Got an App for That. NASA. Retrieved 22 February 2014.

61. Kuhn, M. A.; Povich, M. S.; Luhman, K. L.; Getman, K. V.; Busk, H. A.; Feigelson, E. D.; et al. (2013). The Massive Young Star-Forming Complex Study in Infrared and X-Ray: Mid-infrared Observations and Catalogs. Astrophysical Journal Supplement. **209** *(2): 29. Bibcode*:2013ApJS..209...29K. doi:10.1088/0067-0049/209/2/29.

62. Carey, Bjorn (October 18, 2005). Life›s Building Blocks ‹Abundant in Space›. Space.com. Retrieved March 3, 2014.

63. Hudgins, D. M.; Bauschlicher, C. W. Jr.; Allamandola, L. J. (2005). Variations in the Peak Position of the 6.2 *μm* Interstellar Emission Feature: A Tracer of N in the Interstellar Polycyclic Aromatic Hydrocarbon Population. Astrophysical Journal. **632** *(1): 316–32. Bibcode*:2005ApJ...632..316H. doi:10.1086/432495. Retrieved March 3, 2014.

64. Allamandola, Louis; et al. (April 13, 2011). Cosmic Distribution of Chemical Complexity. NASA. Retrieved March 3, 2014.

65. Clavin, Whitney (February 10, 2015). Why Comets Are Like Deep Fried Ice Cream. NASA. Retrieved February 10, 2015.

66. Battersby, S. (2004). Space molecules point to organic origins. New Scientist. Retrieved 2009-12-11.

67. Mulas, G.; Malloci, G.; Joblin, C.; Toublanc, D. (2006). Estimated IR and phosphorescence emission fluxes for specific polycyclic aromatic hydrocarbons in the Red Rectangle. Astronomy and Astrophysics. **446** *(2): 537–49.* arXiv:astro-ph/0509586. Bibcode:2006A&A...446..537M. doi:10.1051/0004-6361:20053738.

68. Staff (28 July 2010). Bright Lights, Green City. NASA. Retrieved 13 June 2014.

69. García-Hernández, D. A.; Manchado, A.; García-Lario, P.; Stanghellini, L.; et al. (2010). Formation Of Fullerenes In H-Containing Planatary Nebulae. The Astrophysical Journal Letters. **724** *(1): L39–43.* arXiv:1009.4357. Bibcode:2010ApJ...724L..39G. doi:10.1088/2041-8205/724/1/L39.

70. Atkinson, N. (2010-10-27). Buckyballs Could Be Plentiful in the Universe. Universe Today. Retrieved 2010-10-28.

71. Staff (September 20, 2012). NASA Cooks Up Icy Organics to Mimic Life›s Origins. Space.com. Retrieved September 22, 2012.

72. Gudipati, M. S.; Yang, R. (2012). In-Situ Probing Of Radiation-Induced Processing Of Organics In Astrophysical Ice Analogs—Novel Laser Desorption Laser Ionization Time-Of-Flight Mass Spectroscopic Studies (PDF). The Astrophysical Journal Letters. **756** *(1): L24. Bibcode*:2012ApJ...756L..24G. doi:10.1088/2041-8205/756/1/L24.

73. López-Puertas, Manuel (June 6, 2013). PAH›s in Titan›s Upper Atmosphere. CSIC. Retrieved June 6, 2013.

CHAPTER 7

Chemical Bonding

Chapter Contents

- **Introduction**
- **Valence Bond Theory: Lewis Dot Structures, The Octet Rule, Formal Charge, Resonance, and the Isoelectronic Principle**
- **The Shapes of Molecules (Vsepr Theory) and Orbital Hybridization**
- **Bond Polarity and Bond Strength**

Chemical bonding happens when two or more atoms join together to form a molecule. It is a general principle in science that all systems will try to reach their lowest energy level, and chemical bonding will only take place when a molecule can form that has less energy than its uncombined atoms. The three main types of bond are ionic, covalent, and metallic. These all involve electrons moving between atoms in various ways. Another, much weaker, type is the hydrogen bond.

INTRODUCTION

Molecules (and extended solids) are built from atoms that form chemical bonds. Theories of bonding seek to explain why molecules and solids form, what their structures are, why some are more stable than others, and how they react. Quantum mechanics gives us the most realistic picture of chemical bonding via molecular orbital (MO) theory. However, the MO description of bonding is conceptually difficult and mathematically intensive. This chapter will review less rigorous (but still useful) models such as Lewis dot structures and valence shell electron-pair repulsion (VSEPR) theory. When combined with a qualitative quantum mechanical description of bonding through the concepts of orbital hybridization and resonance, these simple models can help us understand a great deal about the structures, stabilities, and reactions of inorganic molecules.

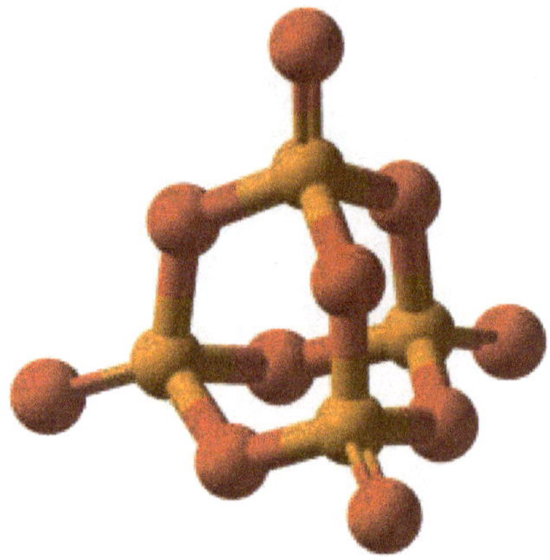

Ball-and-stick representation of chemical bonding in the molecules P_4O_{10} and P_4S_{10}.

The theory of chemical bonding has a long history, dating back to ancient Greece and the atomists Democritus, Leucippus, and the Epicureans. They postulated the existence of immutable atoms moving through the void, and envisioned the physical properties of materials as arising from the kinds and shapes of atoms. In his epic poem *De rerum natura* (On the Nature of Things), the Roman poet Lucretius (c. 99 BC – c. 50 BC), drawing on his Epicurean beliefs, describes some atoms and chemical bonding in the following way:

What seems to us the hardened and condensed
Must be of atoms among themselves more hooked,
Be held compacted deep within, as ‹twere
By branch-like atoms -- of which sort the chief
Are diamond stones, despisers of all blows
And stalwart flint and strength of solid iron
And brazen bars, which, budging hard in locks,
Do grate and scream. But what are liquid, formed
Of fluid body, they indeed must be
Of elements more smooth and round – because
Their globules severally will not cohere.

Lucretius' poem is enjoyable reading and contains some remarkable insights into the microscopic world, given the tools available at the time. Modern analytical methods show that he was off base with his ideas about hooks and spheres, however. We will revisit the nature of chemical bonding in the substances Lucretius mentions (diamond, silicates, iron, brass, and water) in the context of modern chemical theories to understand why they have the special properties they do.

VALENCE BOND THEORY: LEWIS DOT STRUCTURES, THE OCTET RULE, FORMAL CHARGE, RESONANCE, AND THE ISOELECTRONIC PRINCIPLE

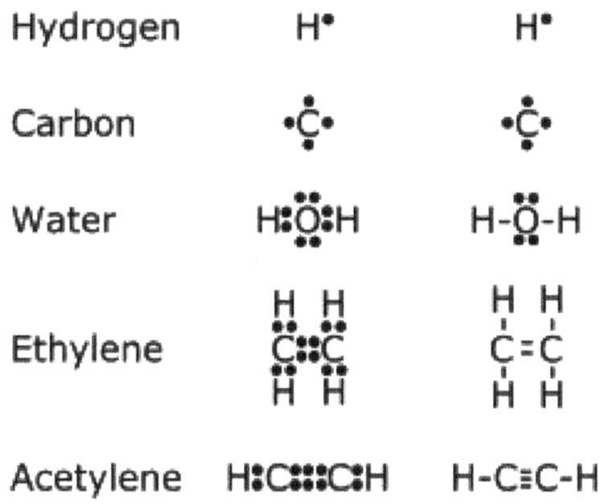

Fig. 1.1.1. Examples of Lewis dot diagrams showing chemical bonds between carbon, hydrogen, and oxygen, and non-bonding electron pairs.

Atoms and Molecules

Atomism, because it was dismissed by Aristotle, enjoyed a long sleep in scientific discourse until it was reconsidered by Galileo, Decartes, and Gassendi in the 1600's. Dalton postulated the modern atomic theory in 1808 based on his observation that elements such as hydrogen and oxygen combined in specific ratios (the Law of Definite Proportions), but the atomic theory remained contentious throughout most of the 19th century. Thompson, Rutherford, Bohr, and others around the turn of the 20th century established that matter was indeed composed of atoms that contained heavy nuclei and light electrons, and that atoms could exist in excited states that could be interpreted as excitations of their electrons to different energy levels. However the atomic theory did not provide a ready explanation for the bonded states of atoms in molecules.

In 1916, still more than a decade before modern quantum theory would adequately describe the shapes of atomic orbitals, Lewis proposed the octet theory based on the empirically observed rules of valence, i.e., the combining ratios of atoms in molecules. This theory, in hindsight, can be rationalized for s- and p-block elements by observing that main-group atoms can use their four valence orbitals (s, p_x, p_y, and p_z) to accomodate up to eight electrons, some or all of which may be shared with other atoms. In Lewis' model, the valence electrons of an atom were situated at the corners of a cube, and the cubes could share edges or faces to complete their octets. Lewis developed a shorthand notation for these structures based on dots that represented the valence electrons, as illustrated in Fig. 1.1.1. A pair of electrons shared between atoms constitutes a chemical bond, and can also be represented as a line joining the atoms. Four electrons shared between atoms, represented by two lines, is a double bond, and so forth. Any pairs of electrons not involved in bonding form "lone pairs" that belong to one atom only and are thus not involved in bonding.

The Lewis picture is powerful in its simplicity. It can be readily used to rationalize or predict the combining ratios of atoms, to sort plausible and unlikely structures for molecules (including large ones), and to rationalize the acid-base properties of many molecules. It is important to remember that the model is built on a shaky, pre-quantum mechanical description of atoms and so with it, we will make mistakes. Nevertheless it is useful to see how far we can go with the Lewis model.

Constructing the octet Valence Bond Picture for a Molecule

We can construct an octet picture for any molecule using the **N-V method:**

Number of electrons *needed* to make separate octets around all atoms = **N**

(N = 8 for non-H atoms, N=2 for H)

Number of *valence* electrons = **V** (add up using group numbers; don't forget to add or subtract for charges on ions)

Number of *shared* electrons = **S** = N-V, number of bonds = S/2

Fill in lone pairs everywhere else to complete octets

Example: nitrate anion, NO_3^-

N = 4 atoms x 8 electrons = 32

V = 5 + 6 + 6 + 6 + 1 = 24 (N + O + O + O + charge)

S = 32-24 = 8 shared electrons = **4 bonds**

The number of lone pairs is obtained by difference: (V-S)/2 = (24-8)/2 = 8

Now draw the molecule, starting with the four bonds and adding eight lone pairs to complete the octets:

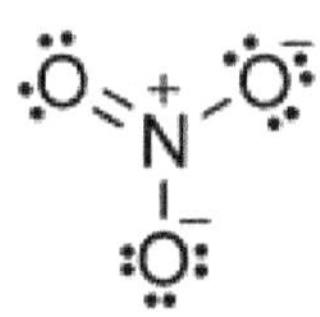

Count the shared and unshared electrons around each atom in the nitrate ion and you will see that all atoms are **octet** and that the total number of valence electrons (V) is 24. If you do the N-V calculation according to the rules above, you should always get an octet structure.

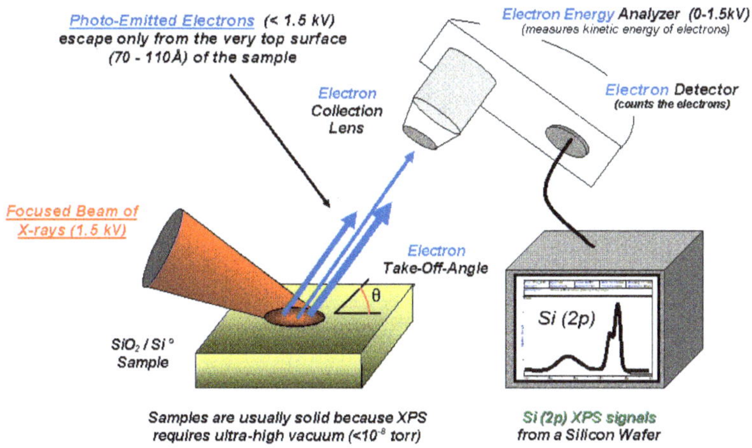

The charges on atoms in a molecule (or an extended solid such as SiO_2) can be estimated from X-ray photoelectron spectra (XPS). Energetic X-rays ionize atoms by kicking out electrons from their core orbitals. By conservation of energy, the kinetic energy (KE) of an emitted electron is related to its binding energy (BE) and the energy of the incoming X-ray photon (hν) by BE = hν - KE. Electrons are harder to remove from positively charged atoms (higher BE), and easier to remove from negatively charged atoms (lower BE), relative to the neutral element.

The **formal charge** distribution is assigned by dividing the shared (bonding) electrons equally between atoms. Thus, the singly bonded O atoms each possess 7 electrons, and because O is in group 6, their formal charge is -1. The doubly bonded O has a formal charge of zero. The N atom has a formal charge of +1 because it "owns" 4 valence electrons and is in group 5.

Note that the **formal charge** is **not** the same thing as the **oxidation number** (or **oxidation state**). In the nitrate ion, the oxidation state of nitrogen is +5 and the oxidation state of oxygen is -2. The formal charge is typically closer to the "real" charge on the atom (as measured, e.g., by X-ray photoelectron spectroscopy). Oxidation states are a useful bookkeeping device for keeping track of oxidation-reduction reactions. Like oxidation states, the formal charges on the atoms in a molecule or ion must *add up to its overall charge*.

We can similarly draw the Lewis structure for ammonia as shown below:

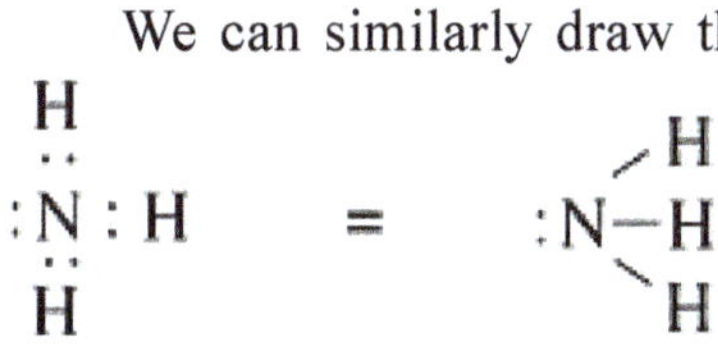

When we divide the shared electrons equally between the atoms, we see that the N atom has five electrons and each H atom has one. These are the same as their group numbers, and thus, all the formal charges in the ammonia molecule are zero.

Octet structures of the Lewis acid-base adduct NH_3BF_3, the hydronium ion H_3O^+, and the sulfate anion SO_4^{2-} are shown below. In this case (and in many Lewis structures we will draw), we leave off the implied lone pairs around the peripheral atoms. Try calculating the number of bonds in each of these molecules using the N-V method, and fill in the lone pairs that are not explicitly drawn in.

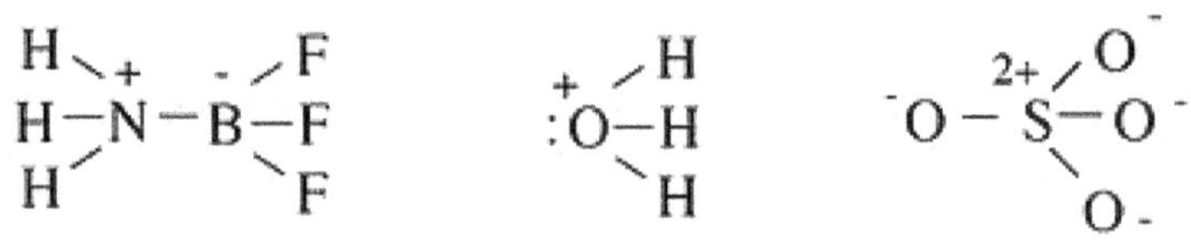

In cases where more than one valence bond structure is possible, we can use formal charges to decide which structures should be more or less stable. The rules are:

- The formal charges on atoms are minimized in stable structures, zero being the best case.
- Negative formal charges should be placed on the most electronegative atom(s).
- Positive formal charges should be placed on the least electronegative atom(s).
- It is unfavorable to place like charges (++ or --) on neighboring atoms in a molecule.

Examples of these rules are shown below for alternative structures of BF_3 and ONF.

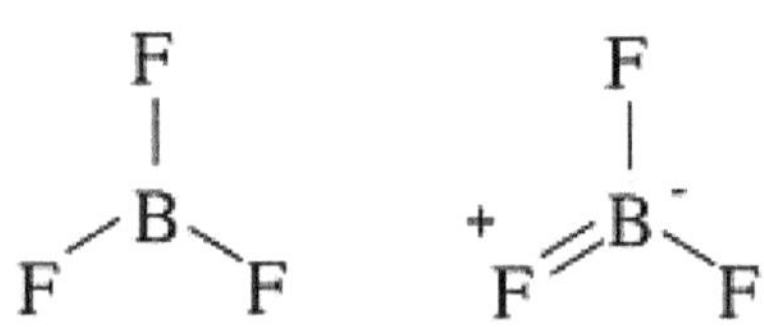

In the BF_3 case, the structure on the left is non-octet, because there are only six electrons (three bonds) in the valence shell of B. Such structures are said to be *electron deficient.* An octet structure (right) can be drawn, but it places a positive formal charge on F, the most electronegative atom in the molecule. Thus, neither structure is completely "happy," but the formal charge rules tell us that the electron-deficient structure on the left is more stable. The electron deficiency of BF_3 makes it a powerful Lewis acid.

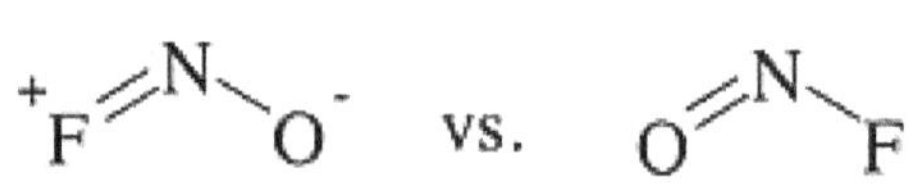

In the case of ONF, the structure on the left is unfavorable for two reasons. First, it places a positive formal charge on F, the most electronegative atom. Second, there is another possible structure (right) that has zero formal charge, which is preferable to one with non-zero charge. This means that the structure on the right, with a double bond between N and O, is more reasonable.

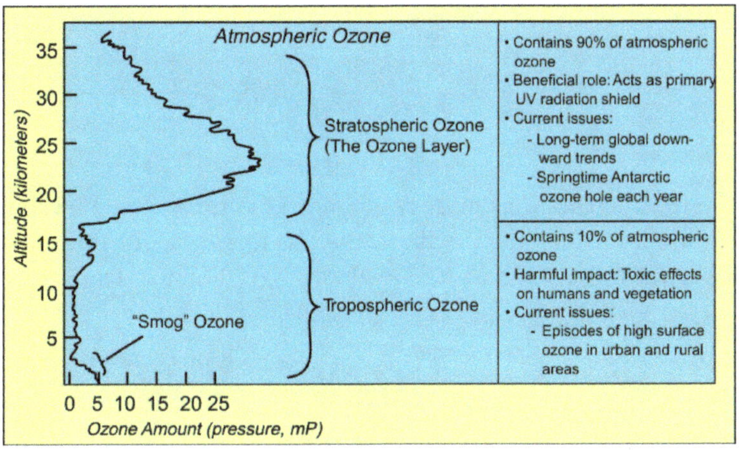

Life outside of the ocean owes its existence to the presence of ozone in the stratosphere. Ozone absorbs the ultraviolet light in the solar spectrum, which otherwise would cause catastrophic damage to DNA and other biological molecules. Because ozone is produced photochemically from O_2, which is itself generated by photosynthesis, the spectroscopic signature of ozone in the atmosphere of distant planets is one possible way to look for extraterrestrial life.

Resonance Structures

The ozone (O_3) molecule has two equivalent octet structures, shown below:

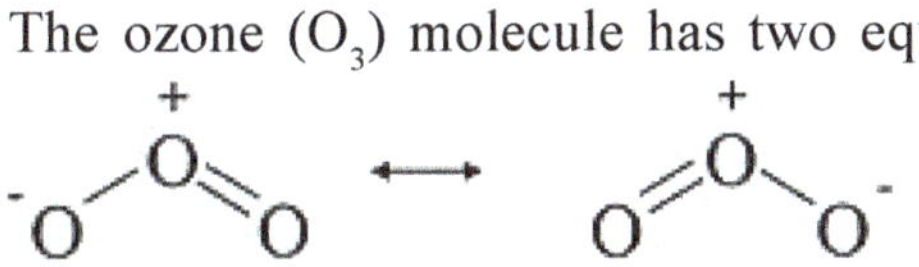

In both cases, the Lewis dot diagram suggests that there are three kinds of oxygen atoms in the molecule, with +1, 0, and -1 formal charges. These structures also suggest that ozone should have one single and one double bond. Experimentally (by electron diffraction), however, we find that the molecule is symmetric, with both O-O bond lengths the same. The real (instantaneous) structure is the average of the two forms, as shown below. In the classical Lewis picture of the molecule, we can rationalize resonance by observing that electrons, being thousands of times lighter than the nuclei of atoms, move very fast on the timescale of molecular vibrations. Thus, in the time it takes for the oxygen atoms to adjust their positions, the electrons can move back and forth many times. In the quantum mechanical MO picture, we will see that resonance involves electrons that are fully delocalized over the atoms in the molecule. We represent resonance structures with a double-headed arrow to signify that the only difference between the Lewis structures is the distribution of electrons.

In the instantaneous structure of ozone, the formal (-) charge is shared between the two terminal oxygen atoms and therefore each one has a formal charge of -1/2. The O-O bonds are the average of a single and a double bond, i.e., each O-O bond order is 1.5.

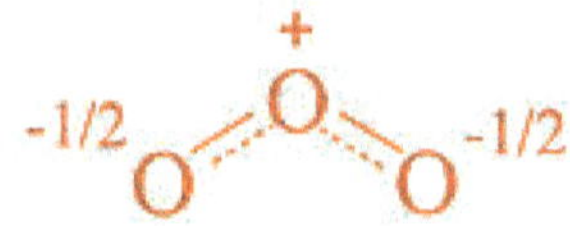

Similarly, the nitrate anion has three resonance structures, and experimentally we find that all the N-O bonds in the ion are the same. In this case, in the averaged structure, each O atom has a formal charge of -2/3 and the N-O bond order is 4/3 (=1.33...).

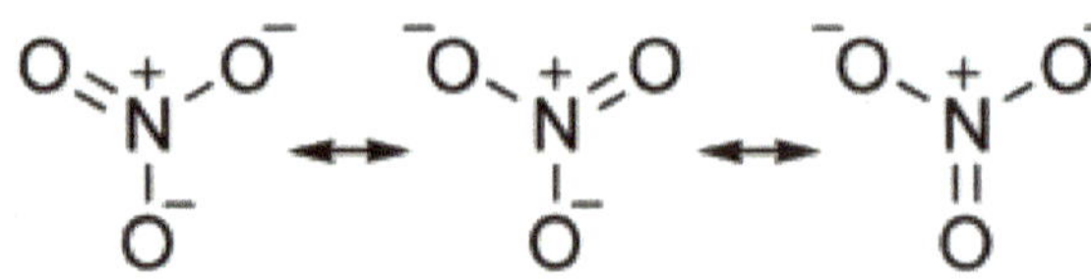

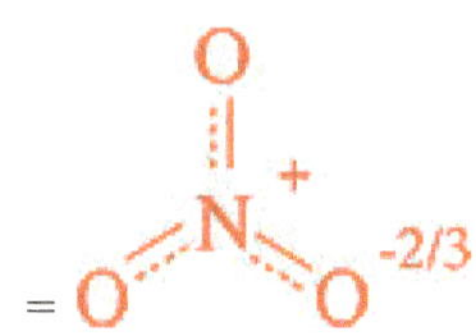

Inequivalent Resonance Structures

The rules of resonance also apply to inequivalent structures, which in general will have different energies from each other. In this case the structure of the molecule represents a weighted average of the low energy structures. A good example is the cyanate ion, OCN^{-}. We can write three inequivalent octet structures for the molecule:

$$^{-}O-C\equiv N \longleftrightarrow O=C=N^{-} \longleftrightarrow {}^{+}O\equiv C-N^{2-}$$

The first two are reasonable resonance structures, although we suspect the one on the left is the best because it puts the negative formal charge on the most electronegative atom. The last one is clearly a bad resonance structure, because the formal charges are high and there is a positive charge on the oxygen atom. The real structure is thus a weighted average of the first two.

No-Bond Resonance

An interesting and useful kind of inequivalent resonance structure is one in which there is a bond order of *zero* between two of the atoms. This concept of **no-bond resonance** is important in understanding the bonding in many halogen- and hydrogen-containing compounds. The idea is illustrated below for the generic molecule X-Y-Z, where Z might be an electronegative halogen atom such as F. By moving the bonding electrons from the Y-Z bond onto the Z atom, and moving a lone pair from X into the X-Y bond, we generate the resonance structure shown on the right in which all atoms remain octet, but there is no bond between Y and Z.

It is important to recognize in this example that the no-bond form is only one resonance structure, and therefore the Z atom is still bound to Y. If the two resonance structures in this example have the same energy, we would expect the X-Y bond order to be 1.5 and the Y-Z bond order to be 0.5. Therefore the Y-Z bond should be longer than it is in a compound where the bond order is one, and it should be relatively easy to break the Y-Z bond.

Some molecular examples of no-bond resonance are shown below. In the molecule ONF_3, the N-F bond is unusually long relative to the N-F bond in NF_3, in which the bond order is 1. This can be explained by the no-bond resonance forms shown on the right. Similarly, in the Lewis acid-base adduct formed by combining BH_3 with CO, we can explain the long B-H bonds using no-bond resonance forms that place a partial positive charge on the H atoms.

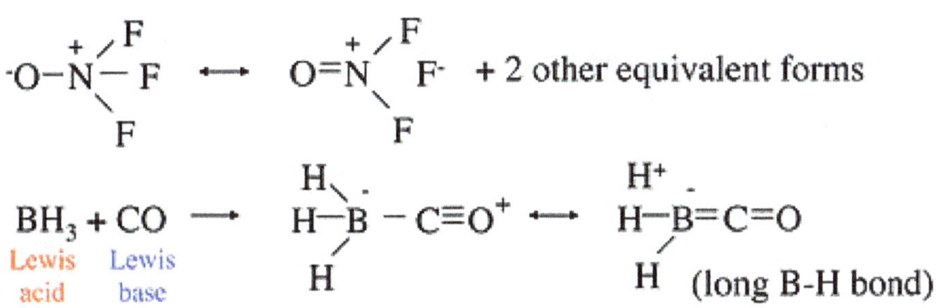

No-bond resonance is often used to provide an octet bonding picture for so-called **hypervalent compounds**, which are compounds that appear to have more than 8 valence electrons in the bonding shell of the central atom. For example, we can consider two different valence bond structures for the triiodide ion, I_3^-, which is formed by reaction of I^- with I_2 in water:

$$I^- + I_2 \rightarrow I_3^- \quad \text{triiodide ion}$$

hypervalent structure: 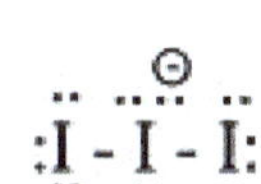

In this structure, the central iodine atom has 10 electrons in its valence shell, in violation of the octet rule. Raman spectra of the triiodide ion show that the I-I bond is weaker than the I-I single bond in I_2, suggesting that this picture is not an accurate description of the bonding. A better representation of I_3^- can be obtained with no-bond resonance structures, as shown below.

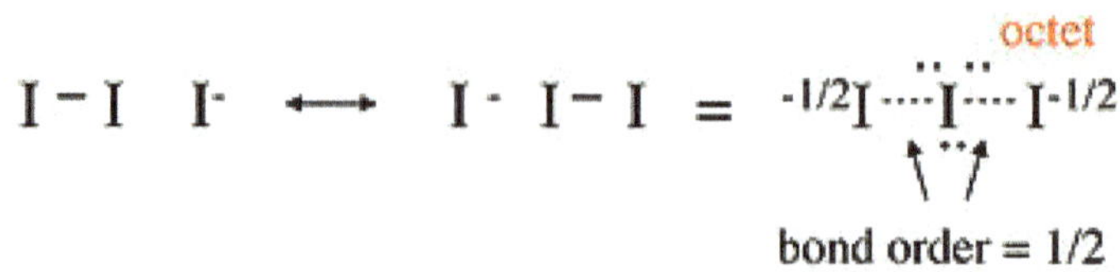

We can draw a similar picture for the XeF_2 molecule, which has the same number of valence electrons as I_3^-. This picture is consistent with XPS data, which show a partial negative charge on the F atoms, as well as vibrational spectra, which show that the Xe-F bond is weaker in XeF_2 than it is in the singly-bonded cation $Xe\text{-}F^+$.

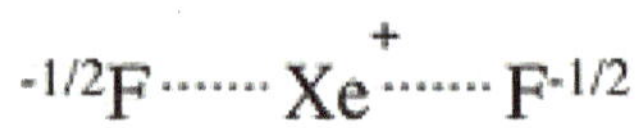

Other well known examples of hypervalent compounds are PF_5, $P(CH_3)_5$, and SF_6, as well as oxyacids such as H_2SO_4 and $HClO_4$. The hypervalent structure is often drawn for these molecules, with the explanation that d-orbitals on the central atom contribute to the bonding in dsp^3 and d^2sp^3 hybrids for 5- and 6-coordinate molecules, respectively. However, realistic molecular orbital calculations show that the phosphorus and sulfur 3d orbitals are too high in energy to contribute significantly to bonding in PF_5 and SF_6. For these molecules, we can use no-bond resonance to make reasonable octet structures that predict polar bonds between the central atom and F.

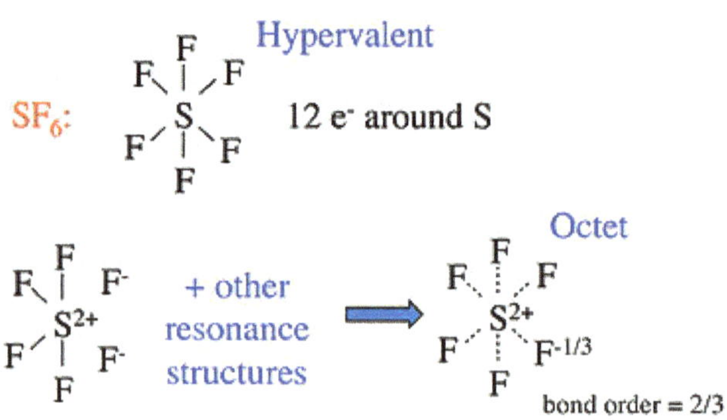

In other cases, such as $P(CH_3)_5$, the octet structure is unreasonable because it suggests a polar bond between P and C, with a partial negative charge on C. Further, in the case of oxyacids such as H_2SO_4 and $HClO_4$, X-ray crystallographic data establish that the S-O and Cl-O bonds are shorter for the oxygen atoms not bonded to hydrogen, which is more consistent with the hypervalent picture:

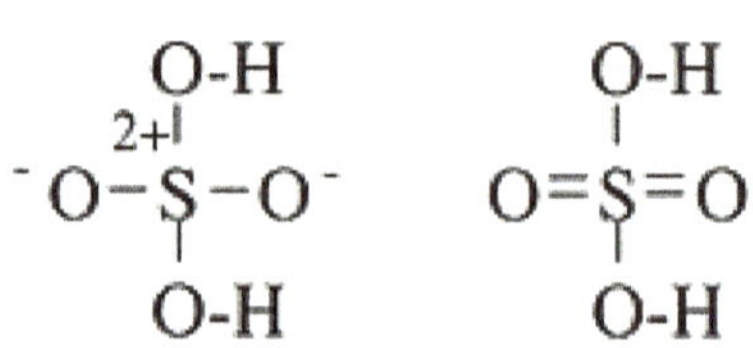

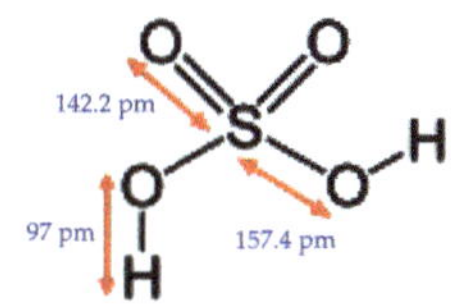

The question of whether hypervalency and the octet rule are really useful descriptions of the bonding in these compounds has been considered in a number of computational studies, which have used increasingly accurate quantum mechanical calculations to determine the number of electrons associated with the central atom. In a 2002 study, Gillespie and Silvi found that the population of the valence shell is greater than eight for compounds with electropositive ligands, such as $P(CH_3)_5$, and less than eight for compounds such as PF_5. They concluded that these valence electron shell populations depend primarily on the coordination numbers and electronegativities of the central atoms and their ligands, and that there is no fundamental difference between the bonding in hypervalent and non-hypervalent (Lewis octet) molecules. This reminds us that the octet rule is not a law of nature, but rather an empirical rule that is useful within certain limits.

Nitrous oxide (N_2O), or laughing gas, is isoelectronic with CO_2. The two molecules have the same shape (linear) and similar physical properties. CO_2 sublimes at -78 °C, whereas N_2O melts at -91 °C and boils at -88 °C.

The Isoelectronic Principle

In calculating the octet structures of molecules using the N-V method, we needed to know only the number of atoms and the number of electrons, not the identities of the atoms themselves. This means that we will get the same answer (and the same set of octet and resonance structures) for *any* molecule or ion that contains (a) the same number of non-hydrogen atoms, and (b) the same total number of valence electrons. Such molecules are said to be *isoelectronic.* This is a powerful conclusion because, once we have determined the electronic structure of one molecule, we can write down the same solution for *all other molecules* that are isoelectronic. For example, we noted above that I_3^- and XeF_2, which both have 22 valence electrons, have the same valence bond structure. We can further expect that isoelectronic molecules will have the same shapes and, very often, similar physical properties.

The four molecules and ions below all contain three non-hydrogen atoms and 16 valence electrons. All of them are linear molecules with two double bonds. The four molecules and ions shown below them all contain four non-hydrogen atoms and 24 valence electrons. These four have a trigonal planar shape. While BF_3 is a member of this isoelectronic series, we do not write it in the resonance form that contains a B=F double bond, because that would put a positive formal charge on F.

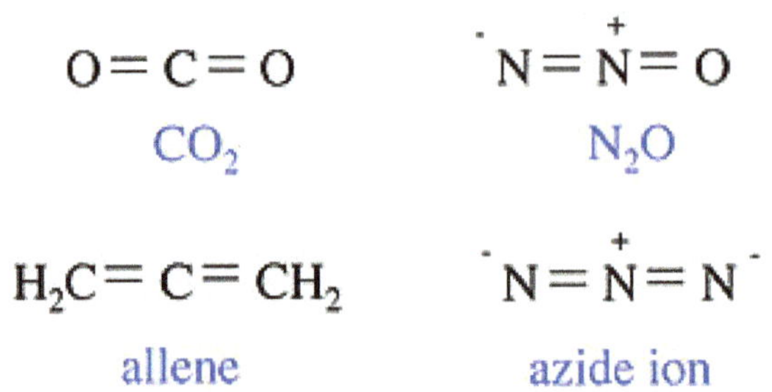

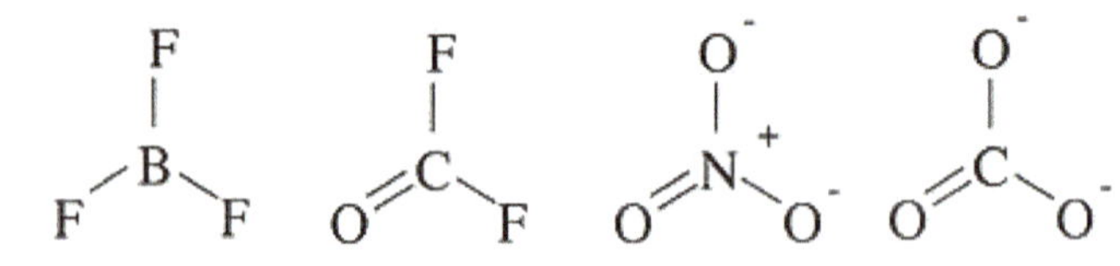

Similarly we can show that CH_4, NH_3, NH_4^+, H_2O, H_3O^+, HF, F^-, and OH^- are all isoelectronic with one non-H atom and eight valence electrons. In the next section we will see that this gives rise to nearly identical electronic shapes for these molecules.

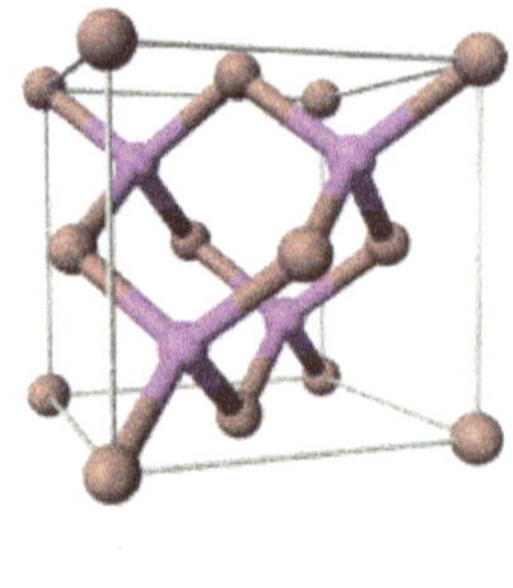

The tetrahedral bonding in crystals of GaAs, CdTe, $CuIn_{1-x}Ga_xSe_2$ and other semiconductors is similar to that in isoelectronic Si and Ge.

Isoelectronic solids. The isoelectronic principle works not just for molecules but for extended solids as well. One technologically important set of isoelectronic solids are the p-block semiconductors. The group 14 element Si is the most widely used semiconductor for electronics, but, as we will discuss later, it is not a good light emitter. Light-emitting diodes (LEDs), which are used in lasers, high efficiency lighting, and display technologies, are made from compounds that are isoelectronic with Si and Ge, especially GaAs, GaP, AlAs, and GaN (all contain four valence electrons per atom). CdTe and $CuIn_{1-x}Ga_xSe_2$(CIGS) are promising solar cell materials that also have the same number of valence electrons per atom. Like Si and Ge, these compounds have tetrahedrally bonded structures in the solid state and absorb light across most of the solar spectrum.

The isoelectronic principle is also a powerful tool in materials research, because it provides guidance about where to look for new materials with similar and perhaps improved properties. For example, the discovery that 8.5% efficient thin layer solar cells could be made with the compound $CsSnI_3$ stimulated the exploration of many isoelectronic ABX_3 compounds with the same perovskite crystal structure. Very recently solar cells in this family based on $CH_3NH_3PbI_{3-x}Cl_x$ have been reported with efficiencies as high as 14-15%.

THE SHAPES OF MOLECULES (VSEPR THEORY) AND ORBITAL HYBRIDIZATION

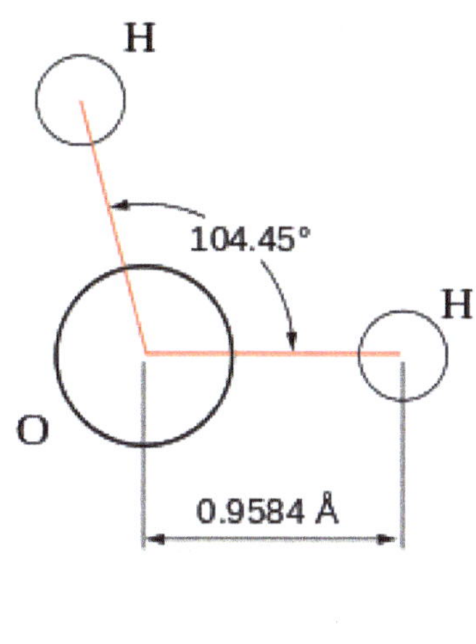

Fig. 1.2.1. Geometry of the water molecule.

The Valence Shell Electron Pair Repulsion (VSEPR) theory is a simple and useful way to predict and rationalize the shapes of molecules. The theory is based on the idea of minimizing the electrostatic repulsion between electron pairs, as first proposed by Sidgwick and Powell in 1940, then generalized by Gillespie and Nyholm in 1957, and then broadly applied over the intervening 50+ years.

To use the VSEPR model, one begins with the Lewis dot picture to determine the number of lone pairs and bonding domains around a central atom. Because VSEPR considers all bonding domains equally (i.e., a single bond, a double bond, and a half bond all count as one electron domain), one can use either an octet or hypervalent structure, provided that the number of lone pairs (which should be the same in both) is calculated correctly. For example, in either the hypervalent or octet structure of the I_3^- ion above, there are three lone pairs on the central I atom and two bonding domains. We then follow these steps to obtain the **electronic geometry**:

- Determine the number of lone pairs on the central atom in the molecule, and add the number of bonded atoms (a.k.a. bonding domains)
- This number (the *steric number*) defines the electronic shape of the molecule by minimizing repulsion. For example a steric number of three gives a trigonal planar electronic shape.
- The angles between electron domains are determined primarily by the electronic geometry (e.g., 109.5° for a steric number of 4, which implies that the electronic shape is a tetrahedron)
- These angles are adjusted by the hierarchy of repulsions: (lone pair - lone pair) > (lone pair - bond) > (bond - bond)

The **molecular geometry** is deduced from the electronic geometry by considering the lone pairs to be present but invisible. The most commonly used methods to determine molecular structure - X-ray diffraction, neutron diffraction, and electron diffraction - have a hard time seeing lone pairs, but they can accurately determine the lengths of bonds between atoms and the bond angles.

The table below gives examples of electronic and molecular shapes for steric numbers between 2 and 9. We are most often concerned with molecules that have steric numbers between 2 and 6.

Bonding electron pairs	**Lone pairs**	**Electron domains (Steric #)**	**Shape**	**Ideal bond angle (example's bond angle)**	**Example**	**Image**
2	0	2	linear	180°	CO_2	
3	0	3	trigonal planar	120°	BF_3	
2	1	3	bent	120° (119°)	SO_2	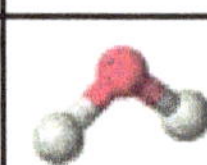
4	0	4	tetrahedral	109.5°	CH_4	
3	1	4	trigonal pyramidal	109.5° (107°)	NH_3	
2	2	4	bent	109.5° (104.5°)	H_2O	

5	0	5	trigonal bipyramidal	90°, 120°, 180°	PCl_5	
4	1	5	seesaw	180°, 120°, 90° (173.1°, 101.6°)	SF_4	
3	2	5	T-shaped	90°, 180° (87.5°, < 180°)	ClF_3	
2	3	5	linear	180°	XeF_2	
6	0	6	octahedral	90°, 180°	SF_6	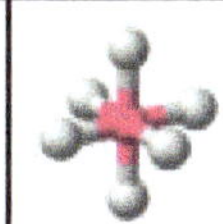
5	1	6	square pyramidal	90° (84.8°), 180°	BrF_5	
4	2	6	square planar	90°, 180°	XeF_4	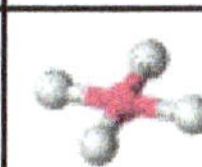
7	0	7	pentagonal bipyramidal	90°, 72°, 180°	IF_7	
6	1	7	pentagonal pyramidal	72°, 90°, 144°	$XeOF_5^-$	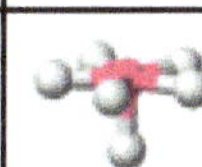
5	2	7	planar pentagonal	72°, 144°	XeF_5^-	
8	0	8	square antiprismatic		XeF_8^{2-}	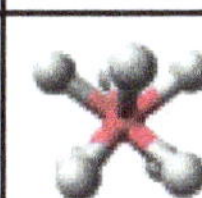
9	0	9	tricapped trigonal prismatic		ReH_9^{2-}	

From the Table, we see that some of the molecules shown as examples have bond angles that depart from the ideal electronic geometry. For example, the H-N-H bond angle in ammonia is 107°, and the H-O-H angle in water is 104.5°. We can rationalize this in terms of the last rule above. The lone pair in ammonia repels the electrons in the N-H bonds more than they repel each other. This lone pair repulsion exerts even more steric influence in the case of water, where there are two lone pairs. Similarly, the axial F-S-F angle in the "seesaw" molecule SF_4 is a few degrees less than 180° because of repulsion by the lone pair in the molecule.

Geometrical Isomers

For some molecules in the Table, we note that there is more than one possible shape that would satisfy the VSEPR rules. For example, the XeF_2 molecule has a steric number of five and a trigonal bipyramidal geometry. There are three possible stereoisomers: one in which the F atoms occupy axial sites, resulting in linear molecule, one in which the F atoms occupy one equatorial and one axial site (resulting in a 90° bond angle), and one in which the F atoms are both on equatorial sites, with a F-Xe-F bond angle of 120°.

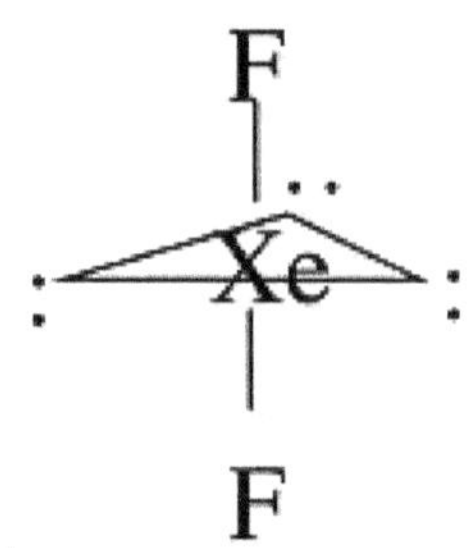

The observed geometry of XeF_2 is linear, which can be rationalized by considering the orbitals that are used to make bonds (or lone pairs) in the axial and equatorial positions. There are four available orbitals, s, p_x, p_y, and p_z. If we choose the z-axis as the axial direction, we can see that the p_x and p_y orbitals lie in the equatorial plane. We assume that the spherical s orbital is shared equally by the five electron domains in the molecule, the two axial bonds share the p_z orbital, and the three equatorial bonds share the p_x and p_y orbitals. We can then calculate the bond orders to axial and equatorial F atoms as follows:

axial: 1/5 s + 1/2 p_z = 0.7 bond (formal charge = -0.3) equatorial: 1/5 s + 1/3 p_x + 1/3 p_y = 0.867 bond (formal charge = -0.133)

Because fluorine is more electronegative than a lone pair, it prefers the axial site where it will have more negative formal charge. In general, by this reasoning, lone pairs and electropositive ligands such as CH_3 will always prefer the equatorial sites in the trigonal bipyramidal geometry. Electronegative ligands such as F will always go to the axial sites.

In the case of the BrF_4^- anion, which is isoelectronic with XeF_4 in the Table, the electronic geometry is octahedral and there are two possible isomers in which the two lone pairs are *cis* or *trans* to each other. In this case, lone pair - lone pair repulsion dominates and we obtain the *trans* arrangement of lone pairs, giving a square planar molecular geometry.

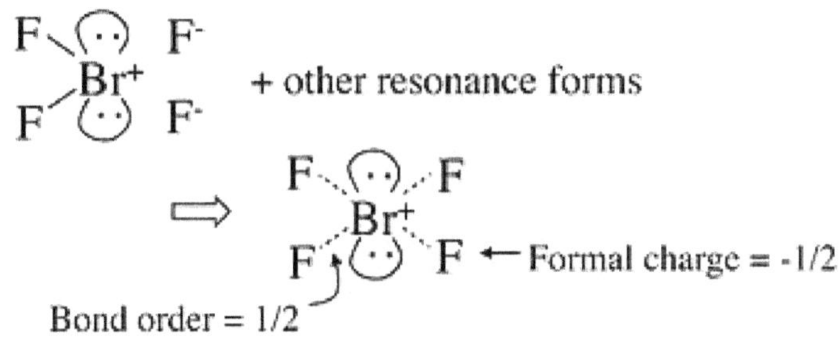

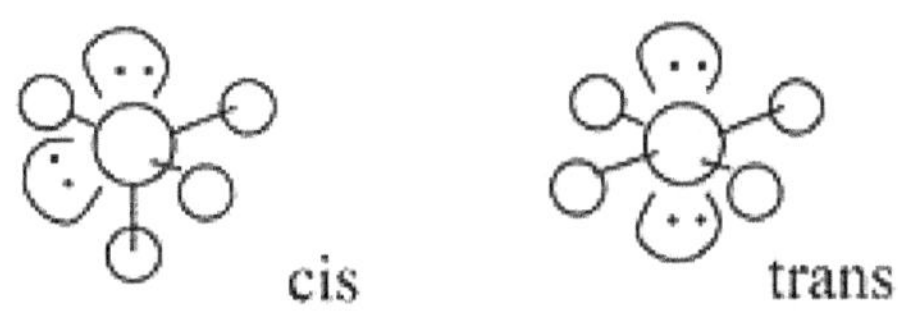

Orbital Hybridization

The observation of molecules in the various electronic shapes shown above is, at first blush, in conflict with our picture of atomic orbitals. For an atom such as oxygen, we know that the 2s orbital is spherical, and that the $2p_x$, $2p_y$, and $2p_z$ orbitals are dumbell-shaped and point along the Cartesian axes (Fig. 1.2.2). The water molecule contains two hydrogen atoms bound to oxygen not at a 90° angle, but at an angle of 104.5°. Given the relative orientations of the atomic orbitals, how do we arrive at angles between electron domains of 104.5°, 120°, and so on?

The atomic orbitals ψ represent solutions to the Schrödinger wave equation, which for hydrogen-like (one-electron) atoms can be written as:

$$E\psi = -\frac{\hbar^2}{2\mu}\nabla^2\psi - \frac{Ze^2}{4\pi\epsilon_0 r}\psi$$

where **Z** is the nuclear charge, *e* is the electron charge, and **r** is the position of the electron. The radial potential term on the right side of the equation is due to the Coulomb interaction, in which ε_0 is the electric constant (permittivity of free space) and

$$\mu = \frac{m_e m_n}{m_e + m_n}$$

is the 2-body reduced mass of the nucleus of mass m_n and the electron of mass m_e. To a good approximation, $\mu \approx m_e$.

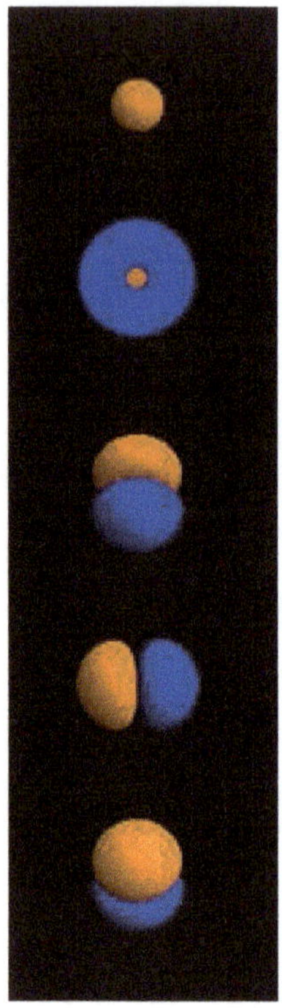

Fig. 1.2.2. The shapes of the first five atomic orbitals: 1s, 2s, $2p_x$, $2p_y$, and $2p_z$. The colors denote the sign of the wave function.

Without going into too much detail about the Schrödinger equation, we can point out some of its most important properties:

- The equation derives from the fact that the total energy (E) is the sum of the kinetic energy (KE) and the potential energy (PE). These three quantities are represented mathematically as **operators** in the equation.
- On the left side of the equation, the total energy operator (E) is a scalar that is multiplied by the wavefunction ψ. ψ is a function of the spatial coordinates (x,y,z) and is related to the probability that the electron is at that point in space.
- The first term on the right side of the equation represents the kinetic energy (KE). The kinetic energy operator is proportional to ∇^2 (the Laplacian operator) which takes the second derivative (with respect to three spatial coordinates) of ψ. Thus, the Schrödinger equation is a **differential equation**.
- The second term on the right side of the equation represents the Coulomb potential (PE), i.e. the attractive energy between the positively charged nucleus and the negatively charged electron.
- The solutions to the Schrödinger equation are a set of **energies** E (which are scalar quantities) and **wavefunctions** (a.k.a. **atomic orbitals**) ψ, which are functions of the spatial coordinates. You will sometimes see the energies referred to as *eigenvalues* and the orbitals as *eigenfunctions*, because mathematically the

Schrödinger equation is an eigenfunction-eigenvalue equation. Although ψ is a function of the coordinates, E is not. So an electron in a $2p_z$ orbital has the same total energy E (= PE + KE) no matter where it is in space.

- These E values and their associated wavefunctions ψ are catalogued according to their **quantum numbers** *n, l,* and m_l. That is, there are many solutions to the differential equation, and each solution (ψ(xyz) and E) has a unique set of quantum numbers. Some sets of orbitals are **degenerate**, meaning that they have the same energy (e.g., $2p_x$, $2p_y$, and $2p_z$).
- The solutions ψ(xyz) to the Schrödinger equation (e.g., the 1s, 2s, $2p_x$, $2p_y$, and $2p_z$ orbitals) represent **probability amplitudes** for finding the electron at a particular point (x,y,z) in space. A probability amplitude can have either + **or - sign**. We typically represent the different signs by shading or by + and - signs on the two lobes of a 2p orbital.
- The square of the probability amplitude, ψ^2, is always a positive number and represents the **probability** of finding the electron at a point x,y,z in space. Because the total probability of finding the electron somewhere is 1, the wavefunction must be **normalized** so that the integral of ψ^2 over the spatial coordinates (from $-\infty$ to $+\infty$) is 1.
- The solutions to the Schrödinger equation are **orthogonal**, meaning that the product of any two (integrated over all space) is zero. For example the product of the 2s and $2p_x$ orbitals, integrated over the spatial coordinates from $-\infty$ to $+\infty$, is zero.

Orbital hybridization involves making *linear combinations* of the atomic orbitals that are solutions to the Schrödinger equation. Mathematically, this is justified by recognizing that the Schrödinger equation is a *linear* differential equation. As such, any sum of solutions to the Schrödinger equation is also a valid solution. However, we still impose the constraint that our hybrid orbitals must be **orthogonal** and **normalized**.

Rules for orbital hybridization:

- Add and subtract atomic orbitals to get hybrid orbitals.
- We get the same number of orbitals out as we put in.
- The energy of a hybrid orbital is the weighted average of the atomic orbitals that make it up.
- The coefficients are determined by the constraints that the hybrid orbitals must be orthogonal and normalized.

For **sp hybridization**, as in the BeF_2 or CO_2 molecule, we make two linear combinations of the 2s and $2p_z$ orbitals (assigning z as the axis of the Be-F bond):

$$\psi_1 = \frac{1}{\sqrt{2}}(2s) + \frac{1}{\sqrt{2}}(2p_z)$$
$$\psi_2 = \frac{1}{\sqrt{2}}(2s) - \frac{1}{\sqrt{2}}(2p_z)$$

Here we have simply added and subtracted the 2s and 2pz orbitals; we leave it as an exercise for the interested student to show that both orbitals are normalized (i.e., $\psi_1^2 = \psi_2^2 = 1$) and orthogonal ($\psi_1\psi_1 = 0$).

What this means physically is explained in Fig. 1.2.3. By combining the 2s and $2p_z$ orbitals we have created two new orbitals with large lobes (high electron probability) pointing along the z-axis. These two orbitals are degenerate and have an energy that is halfway between the energy of the 2s and $2p_z$ orbitals.

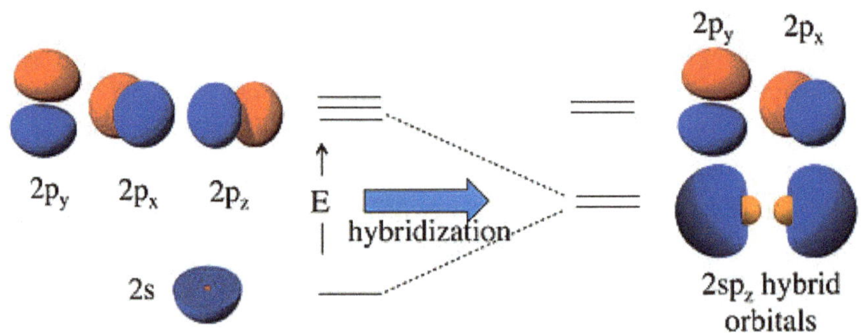

Fig. 1.2.3. Linear combinations of the 2s and $2p_z$ atomic orbitals make two $2sp_z$ hybrids. The $2p_x$ and $2p_y$ orbitals are unchanged.

For an isolated Be atom, which has two valence electrons, the lowest energy state would have two electrons spin-paired in the 2s orbital. However, these electrons would not be available for bonding. By **promoting** these electrons to the degenerate $2sp_z$ hybrid orbitals, they become unpaired and are prepared for bonding to the F atoms in BeF_2. This will occur if the **bonding energy** (in the promoted state) exceeds the promotion energy. The **overall bonding energy**, i.e., the energy released by combining a Be atom in its ground state with two F atoms, is the difference between the bonding and promotion energies.

We can similarly construct **sp^2 hybrids** (e.g., for the BF_3 molecule or the NO_3^- anion) from one 2s and two 2p atomic orbitals. Taking the plane of the molecule as the xy plane, we obtain three hybrid orbitals at 120° to each other. The three hybrids are:

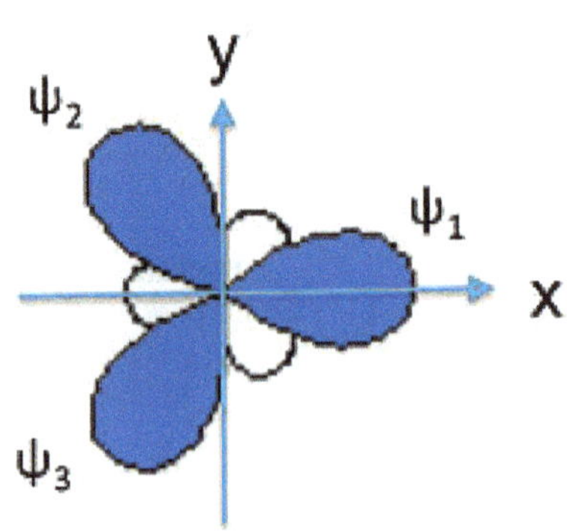

$$\psi_1 = \frac{1}{\sqrt{3}}(2s) + \frac{\sqrt{2}}{\sqrt{3}}(2p_x)$$
$$\psi_2 = \frac{1}{\sqrt{3}}(2s) - \frac{1}{\sqrt{6}}(2p_x) + \frac{1}{\sqrt{2}}(2p_y)$$
$$\psi_3 = \frac{1}{\sqrt{3}}(2s) - \frac{1}{\sqrt{6}}(2p_x) - \frac{1}{\sqrt{2}}(2p_y)$$

These orbitals are again degenerate and their energy is the weighted average of the energies of the 2s, $2p_x$, and $2p_y$ atomic orbitals.

Finally, to make a sp^3 hybrid, as in CH_4, H_2O, etc., we combine all four atomic orbitals to make four degenerate hybrids:

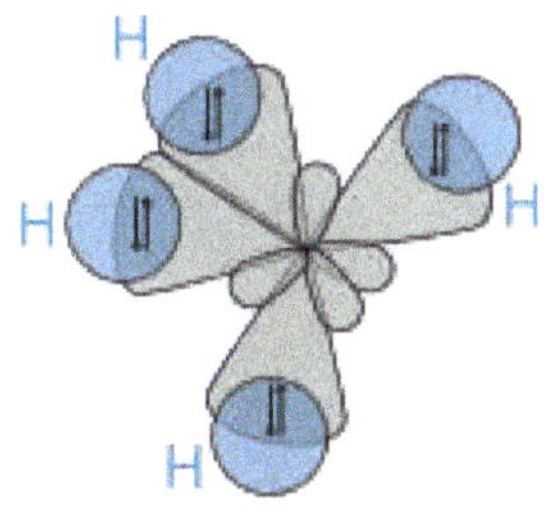

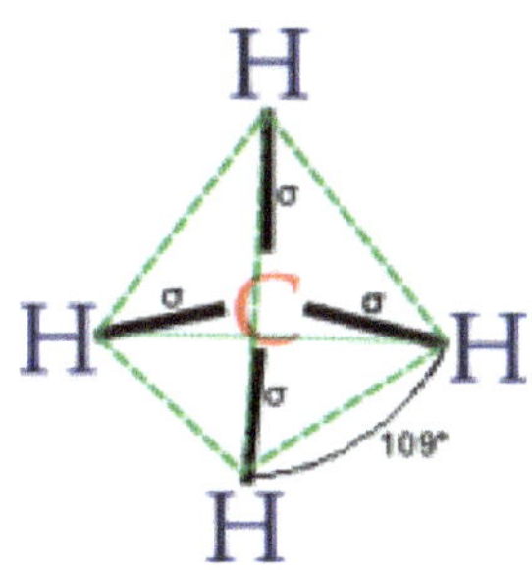

$$\psi_1 = \frac{1}{2}(2s + 2p_x + 2p_y + 2p_z)$$
$$\psi_2 = \frac{1}{2}(2s - 2p_x - 2p_y + 2p_z)$$
$$\psi_3 = \frac{1}{2}(2s + 2p_x - 2p_y - 2p_z)$$
$$\psi_4 = \frac{1}{2}(2s - 2p_x + 2p_y - 2p_z)$$

The lobes of the **sp^3 hybrid** orbitals point towards the vertices of a tetrahedron (or alternate corners of a cube), consistent with the tetrahedral bond angle in CH_4 and the nearly tetrahedral angles in NH_3 and H_2O. Similarly, we can show that we can construct the trigonal bipyramidal electronic shape by making sp and sp^2 hybrids, and the octahedral geometry from three sets of sp hybrids. The picture that emerges from this is that the atomic orbitals can hybridize as required by the shape that best minimizes electron pair repulsions.

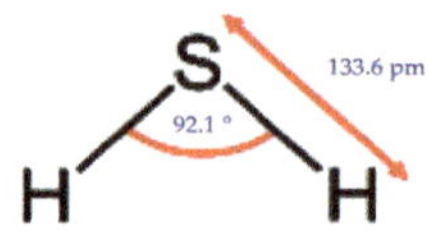

Interestingly however, the bond angles in PH_3, H_2S and H_2Se are close to 90°, suggesting that P, S, and Se primarily use their p-orbitals in bonding to H in these molecules. This is consistent with the fact that the energy difference between s and p orbitals stays roughly constant going down the periodic table, but the bond energy *decreases* as the valence electrons get farther away from the nucleus. In compounds of elements in the 3rd, 4th, and 5th rows of the periodic table, there thus is a *decreasing tendency to use s-p orbital hybrids in bonding*. For these heavier elements, the bonding energy is not enough to offset the energy needed to promote the s electrons to s-p hybrid orbitals.

BOND POLARITY AND BOND STRENGTH

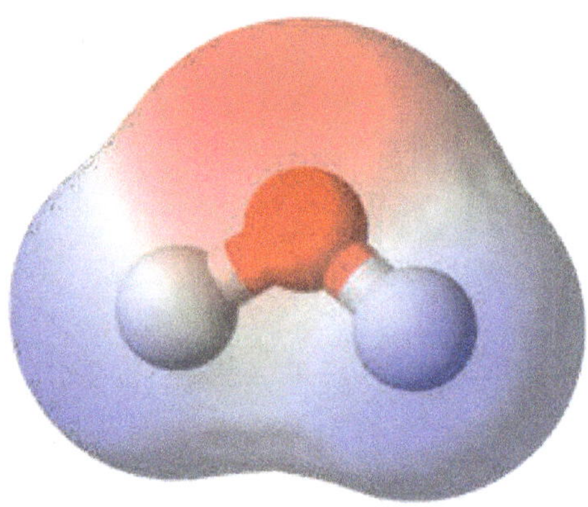

Fig. 1.3.1. A water molecule, a commonly used example of polarity. The two charges are present with a negative charge in the middle (red), and a positive charge at the ends (blue).

Linus Pauling introduced the concept of electronegativity 1932 in order to explain the extra stability of molecules with polar bonds. The electronegativity of an atom, represented by the Greek letter χ, can be defined as the tendency of an atom to draw electrons to itself in a chemical bond. On the Pauling scale, the electronegativity difference between two atoms A and B was defined in terms of the dissociation energies E_d of the A-A, B-B, and A-B bonds:

$$\chi_A - \chi_B = \sqrt{E_d(AB) - [E_d(AA) + E_d(BB)]/2}$$

where the energies are expressed in electron volts.

This definition, while directly relevant to the strength of chemical bonds, requires thermochemical input data from many compounds, some of which were not available at the time. Mulliken and later Pearson developed a scale of electronegativities based on the average of the electron affinity and ionization energy of the free A and B atoms, which they correlated with thermochemical data and the Pauling scale.

On the Pauling scale, the least electronegative elements are the alkali metals (χ = 0.7-1.0) and the most electronegative are oxygen (3.5) and fluorine (4.0) at the upper right of the periodic table. Carbon and hydrogen have intermediate electronegativities (2.6 and 2.2 on the Pauling scale, respectively). The general trend (see table below) is that electronegativities increase going up and to the right in the periodic table. There are some interesting exceptions to this behavior, most notably two islands of high electronegatvity at the bottom of the transition series, peaking at tungsten (χ = 2.4) and gold (χ = 2.5). The first of these can be explained by the very high metal-metal bond energy of elements such as Mo and W, which can use all six of their valence electrons in bonding. The second however occurs with more weakly bonded noble metals such as Pt and Au, and is responsible for their low position in the activity series,[13] as well as their extraordinary properties as catalysts.

Scale: 0.7 – 3.98

	1	2	3	4	5	6	7	8	9	10	11	12	13	14	15	16	17	18
1	H 2.20																	He
2	Li 0.98	Be 1.57											B 2.04	C 2.55	N 3.04	O 3.44	F 3.98	Ne
3	Na 0.93	Mg 1.31											Al 1.61	Si 1.90	P 2.19	S 2.58	Cl 3.16	Ar
4	K 0.82	Ca 1.00	Sc 1.36	Ti 1.54	V 1.63	Cr 1.66	Mn 1.55	Fe 1.83	Co 1.88	Ni 1.91	Cu 1.90	Zn 1.65	Ga 1.81	Ge 2.01	As 2.18	Se 2.55	Br 2.96	Kr
5	Rb 0.82	Sr 0.95	Y 1.22	Zr 1.33	Nb 1.6	Mo 2.16	Tc 2.10	Ru 2.2	Rh 2.28	Pd 2.20	Ag 1.93	Cd 1.69	In 1.78	Sn 1.96	Sb 2.05	Te 2.1	I 2.66	Xe
6	Cs 0.79	Ba 0.89	La 1.10	Hf 1.3	Ta 1.5	W 1.7	Re 1.9	Os 2.2	Ir 2.2	Pt 2.2	Au 2.4	Hg 1.9	Tl 1.8	Pb 1.8	Bi 1.9	Po 2.0	At 2.2	Rn
7	Fr 0.7	Ra 0.9	Ac 1.1	Rf	Db	Sg	Bh	Hs	Mt	Ds	Rg	Uub	Uut	Uuq	Uup			

Lanthanides 6	Ce 1.12	Pr 1.13	Nd 1.14	Pm	Sm 1.17	Eu	Gd 1.20	Tb	Dy 1.22	Ho 1.23	Er 1.24	Tm 1.25	Yb	Lu 1.0
Actinides 7	Th 1.3	Pa 1.5	U 1.7	Np 1.3	Pu 1.3	Am	Cm	Bk	Cf	Es	Fm	Md	No	Lr

Table of Pauling electronegativities

The polarity of bonds is determined by electronegativity differences. As a guideline we define bonds as:

- **ionic** if $\Delta\chi > 2.0$
- **polar** if $2.0 > \Delta\chi > 0.5$

- **nonpolar** if $0.5 > \Delta\chi$

The polarity of bonds helps us understand non-covalent forces between molecules, such as hydrogen bonding and dipole-dipole interactions. It also helps us interpret the reactivity of molecules. For example, the Si-H bond ($\chi_{Si} = 1.8$, $\chi_H = 2.1$) is more hydride-like than the C-H bond ($\chi_C = 2.5$, $\chi_H = 2.1$). Therefore silanes react with acids to make H_2, whereas phosphines ($\chi_P = 2.1$) and hydrocarbons do not. Similarly, electrophilic substitution reactions occur more readily on Si-H and P-H compounds than they do on C-H compounds.

There is also a correlation between the strength of a chemical bond and the bond length, longer bonds being weaker because of weaker orbital overlap. Pauling introduced an empirical formula relating bond length to bond strength. For a given pair of atoms (for example, two carbon atoms):

$$D(n) = D(1) - 0.6 \log_{10}(n)$$

where D(n) represents the bond length in Å and n is the bond order. D(1) in this case would be the length of a C-C single bond, which we can obtain from the average bond length in alkanes (1.54 Å). Using this formula we can predict that the bond lengths in ethylene (C=C double bond) and acetylene (C≡C triple bond) should be 1.36 and 1.25 Å, respectively, which are close to the experimental values of 1.33 and 1.20 Å. In a related form the Pauling formula can be used to calculate bond lengths when the single bond length D(1) is not available:

$$D(n) = D(m) - 0.6 \log_{10}(n/m)$$

Here n and m represent two different bond orders between the same kinds of atoms. This tells, for example, that the difference in length between a triple and double bond, D(2)-D(3), should be $- 0.6 \log_{10}(2/3) = 0.11$ Å.

Some bond lengths and bond energies are anomalous. For example, the F-F bond length in F_2 is 1.43 Å, which is 0.15 Å *longer* than twice the covalent radius of the F atom (0.64 Å). The F-F bond is also quite weak (bond dissociation energy = 38 kcal/mol) relative to the Cl-Cl bond (58 kcal/mol). By putting the extra bond length into the Pauling formula, we calculate that the bond order in the F_2 molecule is only 0.6, i.e., substantially weaker than a F-F single bond. The physical reason for this is that the F-F bond is stretched by repulsion of the lone pairs on the F atoms. This crowding is caused by the fact that the [He] $1s^2$ core orbital, as well as the valence orbitals of the fluorine atoms, are contracted by the high nuclear charge. The Cl_2 atom, with its larger [Ne] ($1s^22s^22p^6$) core, in contrast, has a "normal" single bond length (1.98 Å) that is twice the covalent radius of the Cl atom (0.99 Å). A similar lone pair repulsion effect explains the anomalously long and weak N-N and O-O single bonds in hydrazine (H_2N-NH_2) and hydrogen peroxide (HO-OH), which are both highly reactive molecules.

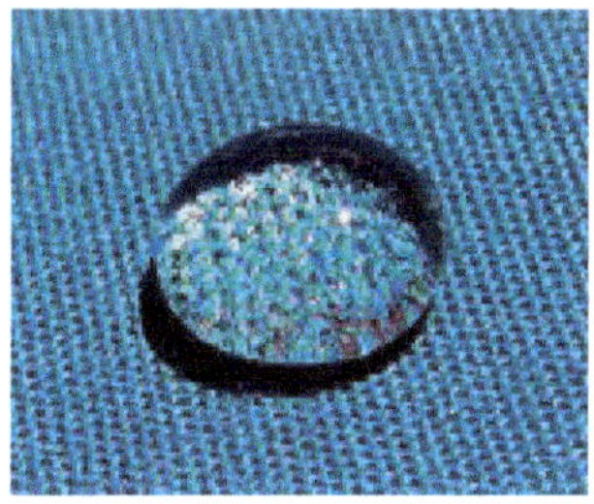

Fabric treated with fluorosurfactant

The weak bond in F_2 is responsible for the high electronegativity of fluorine, as well as the legendary reactivity of elemental fluorine gas, which reacts explosively with hydrogen and powdered metals. Because of the instability of elemental fluorine and the polar nature of its bonds with more electropositive elements, fluorine compounds tend to be very stable. For example, the noble gases Xe and Kr react with fluorine to make covalent compounds, whereas other halogens do not react. Fluorocarbon compounds contain strong C-F bonds and have high thermal and chemical stability. Perfluorocarbons such as Teflon (poly(tetrafluoroethylene), -(CF_2CF_2)n-, PTFE) are also highly hydrophobic. The extraordinary hydrophobicity of perfluorocarbons arises from the fact that -CF_2- and -CF_3 groups are "fatter" than -CH_2- and -CH_3 groups; dissolving them in water is therefore more disruptive to the hydrogen bonding network than is dissolving a hydrocarbon.

REFERENCES

1. G. N. Lewis, "The atom and the molecule," J. Am. Chem. Soc. 1916, 38, 762-785
2. R. J. Gillespie and B. Silvi, The octet rule and hypervalence: two misunderstood concepts, Coord. Chem. Rev. 233-234, 53-62 (2002).
3. I. Chung, B. Lee, J. He, R. P. H. Chang and M. G. Kanatzidis, All-solid-state dye-sensitized solar cells with high efficiency, Nature 485, 486-489 (2012). doi:10.1038/nature11067
4. S.D. Stranks, G. E. Eperon, G. Grancini, C. Menelaou, M. J. P. Alcocer, T. Leijtens, L. M. Herz, A. Petrozza, and H. J. Snaith, Electron-Hole Diffusion Lengths Exceeding 1 Micrometer in an Organometal Trihalide Perovskite Absorber, Science 342, 341-344 (2013). DOI: 10.1126/science.1243982
5. G. Xing, N. Mathews, S. Sun, S. S. Lim, Y. M. Lam, M. Grätzel, S. Mhaisalkar, and T. C. Sum, Long-Range Balanced Electron- and Hole-Transport Lengths in Organic-Inorganic CH3NH3PbI3, Science 342, 344-347 (2013). DOI: 10.1126/science.1243167
6. N. V. Sidgwick and H. M. Powell, Proc. Roy. Soc. A176, 153 (1940),
7. R. J. Gillespie and R. S. Nyholm, Quart. Rev. Chem. Soc., 11, 339 (1957).
8. R. J. Gillespie, "Fifty years of the VSEPR model," Coord. Chem. Rev. 252, 1315-1327 (2008). DOI: 10.1016/j.ccr.2007.07.007

9. Pauling, L. (1932). "The Nature of the Chemical Bond. IV. The Energy of Single Bonds and the Relative Electronegativity of Atoms". *J. Am. Chem. Soc.* 54 (9): 3570–3582. doi:10.1021/ja01348a011.

10. Mulliken, R. S. (1934). A New Electroaffinity Scale; Together with Data on Valence States and on Valence Ionization Potentials and Electron Affinities. *J. Chem. Phys.* 2 (11): 782–793. doi:10.1063/1.1749394. Bibcode: 1934JChPh...2..782M.

11. Mulliken, R. S. (1935). Electronic Structures of Molecules XI. Electroaffinity, Molecular Orbitals and Dipole Moments. J. Chem. Phys. 3 (9): 573–585. doi:10.1063/1.1749731. Bibcode: 1935JChPh...3..573M.

12. Pearson, R. G. (1985). "Absolute electronegativity and absolute hardness of Lewis acids and bases". J. Am. Chem. Soc. 107 (24): 6801. doi:10.1021/ja00310a009.

13. B. Hammer and J. K. Norskov, Why gold is the noblest of all the metals, *Nature* 376, 238 - 240 (2002. doi:10.1038/376238a0

14. V. H. Dalvi and P. J. Rossky, Molecular origins of fluorocarbon hydrophobicity, Proc. Natl. Acad. Sci. USA 107,13603–13607 (2010). DOI: 10.1073/pnas.0915169107

CHAPTER 8

Acid-Base Chemistry

Chapter Contents

- **Introduction**
- **Brønsted and Lewis Acids and Bases**
- **Hard and Soft Acids And Bases**
- **Problems**

Acid-base reactions are chemical reactions that occur between acids and bases. An acid is a substance that, when dissolved in water, has greater hydrogen ion activity than pure water does, while a base is a substance that, when dissolved in water, can accept hydrogen ions. The acidity of a solution is measured based on the pH scale; a substance with a pH less than seven is acidic while a solution with a pH greater than seven is basic. There are many different explanations and forms of acid-base reactions, as they can occur in many different ways and have been studied by a number of different chemists.

INTRODUCTION

Acids and bases are important for a number reasons in inorganic chemistry.

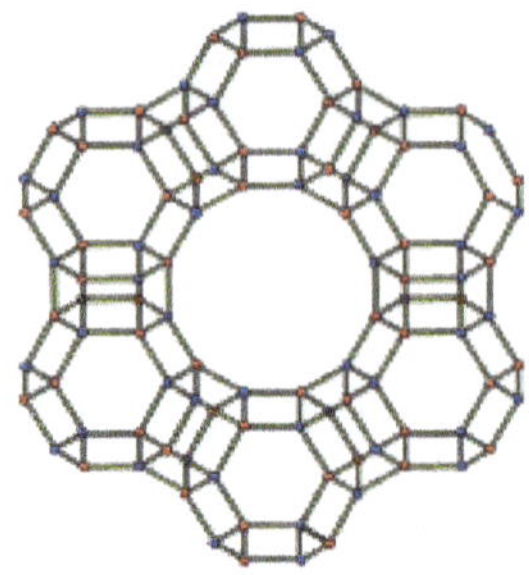

Solid acid catalysts, such as zeolite Y, are used to isomerize hydrocarbons in the processing of crude oil into gasoline.

- Many industrially useful catalytic reactions involve inorganic acids and superacids, such as zeolites, anhydrous hydrogen fluoride, and sulfated zirconia. These acids are sufficiently strong in anhydrous media that they can protonate olefins and alcohols to produce carbocations. Carbocations are key intermediates in the transformations of hydrocarbons.
- Inorganic compounds are sometimes synthesized in strongly acidic or basic media. For example, ternary metal oxides can be synthesized and crystallized in molten NaOH or KOH, which are strongly basic. Organic fluorination reactions are often done in strongly acidic media, such as anhydrous HF. Understanding the familiar chemistry of acids and bases in water helps us understand how these non-aqueous media work.
- The acidic or basic environment of metal ions affects the stability of their oxidation states..
- Transition metal complexes (coordination compounds and organometallic compounds) are essentially Lewis acid-base complexes. We can understand a great deal about their stability and reactivity by considering the acid-base character of metals and ligands.

BRØNSTED AND LEWIS ACIDS AND BASES

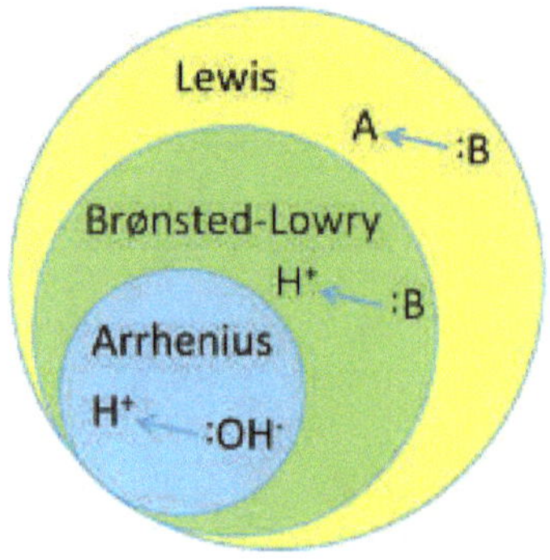

Illustration of the hierarchy of acid-base theories. Arrhenius acids and bases are a sub-class of Bronsted acids and bases, which are themselves a subclass of Lewis acids and bases.

Three theories of acids and bases. There are three major classifications of substances known as acids or bases. The Arrhenius definition states that an acid produces H^+ in solution and a base produces OH^-. This theory was developed by Svante Arrhenius in 1883. Later, two more sophisticated and general theories were proposed. These are the Brønsted-Lowry and the Lewis definitions of acids and bases. The relationship between these theories is illustrated in the figure at the left.

In acid-base chemistry, the hydrogen ion is often called and written as a proton (H^+), but in fact it exists in a variety of hydrated forms that include H_3O^+, $H_5O_2^+$, $H_7O_3^+$. When we write $H^+(aq)$ or $H_3O^+(aq)$ we are referring to the dynamic equilibrium mixtures of these forms of protonated water.

The Arrhenius theory, which is the simplest and least general description of acids and bases, includes acids such as $HClO_4$ and bases such as NaOH or $Mg(OH)_2$. This theory successfully describes how acids and bases react with each other to make water and salts. However, it does not explain why some substances that do not contain hydroxide ions, for example F^- and NO_2^-, can make basic solutions in water. The Brønsted-Lowry definition of acids and bases addresses this problem. In this theory an **acid is a substance that can release a proton** (like in the Arrhenius theory) and a **base is a substance that can accept a proton**. A basic salt such as Na^+F^- generates OH^- ions in water by taking protons from water itself (to make HF):

$$F^-(aq) + H_2O(l) \rightleftharpoons HF(aq) + OH^-$$

When a Brønsted acid dissociates, it increases the concentration of hydrogen ions in the solution, $[H^+]$; conversely, Brønsted bases dissociate by taking a proton from the solvent (water) to generate $[OH^-]$.

Acid dissociation: $HA(aq) \rightleftharpoons A^-(aq) + H^+(aq)$

$$K_a=[A^-][H^+]/[HA]$$

Base dissociation: $B(aq) + H_2O(l) \rightleftharpoons HB^+(aq) + OH^-(aq)$

$$K_b = [HB^+][OH^-]/[B]$$

Conjugate acids and bases. One important consequence of these equilibria is that every **acid** (HA) has a **conjugate base** (A-), and vice-versa. In the base dissociation equilibrium above the conjugate acid of base B is HB^+.

For a given acid or base, these equilibria are linked by the water dissociation equilibrium:

$$H_2O(l) \rightleftharpoons H^+(aq) + OH^-(aq)$$

$$K_w = [H^+][OH^-]$$

for which the equilibrium constant K_w is 1.00×10^{-14} at 25°C. It can be easily shown that the product of the acid and base dissociation constants K_a and K_b is K_w.

Strong and weak acids and bases. Acids and bases that dissociate completely are said to be strong:

$$HClO_4(aq) \rightarrow H^+(aq) + ClO_4^-(aq)$$

$$HBr(aq) \rightarrow H^+(aq) + Br^-(aq)$$

$$CH_3O^-(aq) + H_2O(l) \rightarrow CH_3OH(aq) + OH^-(aq)$$

$NH_2^-(aq) + H_2O(l) \rightarrow NH_3(aq) + OH^-(aq)$

Here the right-handed arrow (→) implies that the reaction goes to completion. That is, a 1.0 M solution of $HClO_4$ in water actually contains 1.0 M $H^+(aq)$ and 1.0 M ClO_4^- (aq), and very little undissociated $HClO_4$.

Conversely, weak acids such as acetic acid (CH_3COOH) and weak bases such as ammonia (NH_3) dissociate only slightly in water - typically a few percent, depending on their concentration and the values of K_a and K_b - and exist mostly as the undissociated molecules.

Antacid tablets typically contain calcium salts of the bicarbonate ion (HCO_3^-), a weak base. Its conjugate acid, carbonic acid (H_2CO_3) is a weak acid. The acid-base equilibrium between carbonic acid and bicarbonate is important in maintaining blood pH.

A common misconception is that strong acids have weak conjugate bases, and that weak acids have strong conjugate bases. It is easy to see that this is incorrect by remembering that $K_aK_b = K_w$. Our definition of a strong acid or base is that $K >> 1$, i.e., that the substance dissociates completely. Our definition of a weak acid or base is $1 > K > K_w$. It follows that if $K_a >> 1$ (strong) then K_b cannot be $> K_w$ (weak).

In fact, **strong acids** such as HCl dissociate to produce **spectator ions** such as Cl^- as conjugate bases, whereas **weak acids** produce **weak conjugate bases**. This is illustrated below for acetic acid and its conjugate base, the acetate anion. Acetic acid is a weak acid ($K_a = 1.8 \times 10^{-5}$) and acetate is a weak base ($K_b = K_w/K_a = 5.6 \times 10^{-10}$)

The strength of a conjugate acid/base varies inversely with the strength or weakness of its parent acid or base. Any acid or base is technically a conjugate acid or conjugate base also; these terms are simply used to identify species in solution (i.e acetic acid is the conjugate acid of the acetate anion, a base, while acetate is the conjugate base of acetic acid, an acid).

Neutral oxyacids (H_2SO_4, H_3PO_4, HNO_3, $HClO_2$, etc.) can be classified as strong or weak following a simple rule first noted by Linus Pauling. If the number of oxygen atoms exceeds the number of hydrogen atoms by two or more, then the acid is strong; otherwise it is weak. For example $HClO_4$ and $HClO_3$, where the difference is 3 and 2, respectively, are both strong acids. HNO_2 and $HClO_2$ are both weak because the difference is 1 in both cases. For weak acids, the relative strength depends on this difference (i.e., $HClO_2$ is a stronger weak acid than HOCl) and on the electronegativity of the central atom (HOCl is stronger than HOI).

Acids that can donate more than one proton are called **polyprotic acids**. For example, sulfuric acid, H_2SO_4, is a strong acid that has a conjugate base that actually happens to be a weak acid itself. This means that every mole of H_2SO_4 in aqueous solution donates more than 1 mole of protons. Carbonic acid (H_2CO_3) and phosphoric acid (H_3PO_4) are weak polyprotic acids. Typically, the sequential pK_a's of polyprotic acid are separated by about 5 pH units, because it becomes progressively more difficult to remove protons as the ion becomes more negatively charged. For example, the three pK_a's of phosphoric acid are 2.15, 7.20, and 12.35.

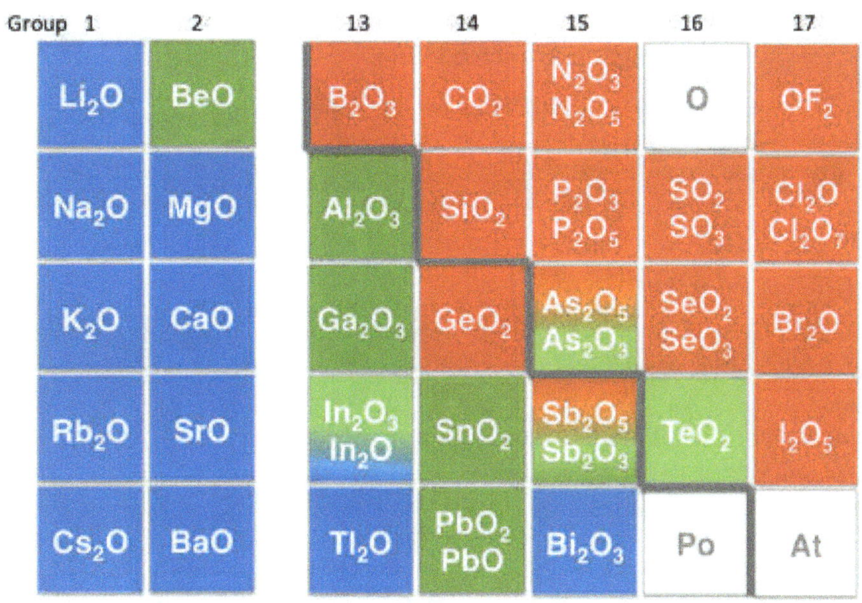

Periodic table showing basic (blue), amphoteric (green) and acidic (red) oxides. The metal-nonmetal boundary is indicated by the gray staircase line.

Amphoteric compounds. Some substances can act either as an acid and as a base. An example is water. H_2O molecules may either donate a hydrogen ion or accept one. This property makes water an **amphoteric** solvent. In the situation where an acid dissociates in solution, water is acting as a base. Conversely, water acts as an acid when bases dissociate. The strongest acid we can make in H_2O is H^+(aq), and the strongest base we can make in H_2O is OH^- (aq).

Other examples of amphoteric compounds are oxides and hydroxides of elements that lie on the border between the metallic and non-metallic elements in the periodic table. For example, aluminum hydroxide ($Al(OH)_3$) is insoluble at neutral pH, but can accept protons in acid to make $[Al(H_2O)_6]^{3+}$ or accept an OH^- ion in base to form $Al(OH)_4^-$ ions. Consequently, aluminum oxide is soluble in acid and in base, but not neutral water. Other examples of amphoteric oxides are BeO, ZnO, Ga_2O_3, Sb_2O_3, and PbO. Increasing the oxidation state of a metal increases the acidity of its oxide by withdrawing electron density from the oxygen atoms. Thus, Sb_2O_5 is acidic, but Sb_2O_3 is amphoteric.

Solvent leveling. Solvent leveling is an effect that occurs when a strong acid is placed in a solvent such as (but not limited to) H_2O. Because strong acids donate their protons to the solvent, the strongest possible acid that can exist is the conjugate acid of the solvent. In aqueous solution, this is H_3O^+. This means that the strength of acids such as HCl and HBr cannot be differentiated in water as they both are dissociated 100% to H_3O^+. In the context of our discussion of conjugate bases above, we would say that both Cl^- and Br^- are spectator ions in water: neither one is a strong enough base to accept a proton from H_3O^+. In order to differentiate the acidities of strong acids such as $HClO_4$ and HCl, or the basicities of strong bases such as CH_3O^- and NH_2^-, we must typically work in non-aqueous solvents, as explained below.

Nonaqueous solutions. The Brønsted theory encompasses any type of solvent that can donate and accept H^+ ions, not just aqueous solutions. The strength of an acid or a base varies depending on the solvent. Non-aqueous acid-base chemistry follows similar rules to those developed for acids and bases in water. For example in liquid ammonia, the solvent autodissociates in the reaction:

$$2NH_3(l) \rightarrow NH_4^+ + NH_2^-$$

This equilibrium is analogous to the autodissociation of water, but has a smaller equilibrium constant ($K \approx 10^{-30}$). It follows by analogy to water that NH_4^+ is the strongest acid and NH_2^- is the strongest base that can exist in liquid ammonia. Because ammonia is a basic solvent, it enhances the acidity and suppresses the basicity of substances dissolved in it. For example, the ammonium ion (NH_4^+) is a weak acid in water ($K_a = 6 \times 10^{-10}$), but it is a strong acid in ammonia. Similarly, acetic acid is weak in water but strong in ammonia. Solvent leveling in fact makes CH_3COOH and NH_4Cl both strong acids in ammonia, where they have equivalent acid strength.

Strong acids that are leveled in water have different acid strengths in acidic solvents such as HF or anhydrous acetic acid. For example, acid dissociation of HX in acetic acid (CH_3COOH) involves protonating the solvent to make its conjugate acid ($CH_3COOH_2^+$) and the X^- anion. Because $CH_3COOH_2^+$ is a stronger acid that H_3O^+, the anion X^- (which is a spectator in water) can become a weak base in CH_3COOH:

$$HX + CH_3COOH \rightleftharpoons CH_3COOH_2^+ + X^-$$

It follows that acidic solvents magnify the Brønsted basicities of substances that cannot accept protons in water. Conversely, basic solvents magnify the acidity of substances that cannot donate a proton to OH^-.

The acidity and basicity of non-aqueous solvents is difficult to quantify precisely, but one good relative measure is the Hammett acidity function, H_o. H_o is defined analogously to pH according to the Henderson-Hasselbach equation:

$$H_o = pK_a + \log([\text{base}]/[\text{conjugate acid}])$$

For non-aqueous solvents, or for acidic or basic compounds in dissolved in solvents that do not themselves dissociate, H_o is a rough measure of the pH of the solvent or compound in question. Anhydrous HF and H_2SO_4 have H_o values of approximately -10 and -12 respectively.

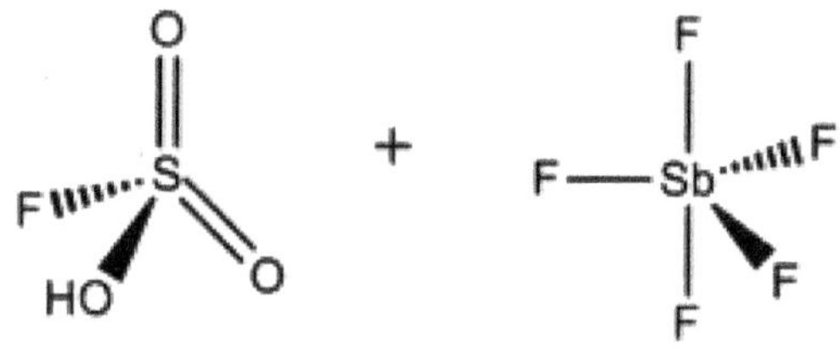

Magic acid is made by mixing FSO_3H and SbF_5. Their reaction generates the $H_2SO_3F^+$ cation, which can protonate hydrocarbons.

Superacids and superbases. The term *superacid* was originally coined by James Bryant Conant in 1927 to describe acids that were stronger than conventional mineral acids.[1] George A. Olah prepared the so-called magic acid, so-named for its ability to attack hydrocarbons, by mixing antimony pentafluoride (SbF_5) and fluorosulfonic acid (FSO_3H). The name was coined after a candle was placed in a sample of magic acid. The candle dissolved, showing the ability of the acid to protonate hydrocarbons, which under aqueous acidic conditions cannot be protonated.

At 140 °C , $FSO_3H–SbF_5$ converts methane into the tertiary-butyl carbocation, a reaction that begins with the protonation of methane:[2]

$CH_4 + H^+ \rightarrow CH+5$

$CH+5 \rightarrow CH+3 + H_2$

$CH+3 + 3\ CH_4 \rightarrow (CH_3)_3C^+ + 3H_2$

Fluoroantimonic acid, $HSbF_6$, can produce solutions with H_0 down to –28.[3] Fluoroantimonic acid is made by combining HF and SbF_5. In this system, HF releases its proton (H^+) concomitant with the binding of F^- by antimony pentafluoride, which (as described below) is a Lewis acid. The resulting anion (SbF−6) is both a weak nucleophile and an extraordinarily weak base.

Superacids are useful in reactions such as the isomerization of alkanes. Industrially, anhydrous acid-exchanged zeolites, which are superacid catalysts, are used on a massive scale to isomerize hydrocarbons in the processing of crude oil to gasoline. Superbases such as lithium diethylamide ($LiNEt_2$), alkyllithium compounds (RLi), and Grignard reagents (RMgX) useful in a broad range of organic reactions. $LiNEt_2$ deprotonates C-H bonds to generate reactive carbanions. RLi and RMgX are powerful nucleophiles.

The use of superbases in nonaqueous media allows us to rank the acidities (and measure the pKa›s) of different classes of molecules. This ranking is particularly important in understanding the reactions of organic molecules. Note that the order of acidities for hydrocarbons is alkynes >> alkenes, aromatics >> alkanes. This ordering has to do with the hybridization of the carbon atom that forms the carbanion. The negatively charged lone pair of the carbanion is stabilized in orbitals that have high s character (e.g., sp vs. sp^2 or sp^3). This is because s orbitals have finite probability density at the nucleus and "feel" the positive nuclear charge (thereby stabilizing the extra negative charge on carbon) more than p orbitals. Resonance effects also stabilize carbanions. Thus, cyclopentadiene is more acidic than even an alkyne because the negative charge is delocalized over the entire (aromatic) $C_5H_5^-$ ring when the C_5H_6 is deprotonated.

name	formula	structural formula	pKa
Methane	CH_4	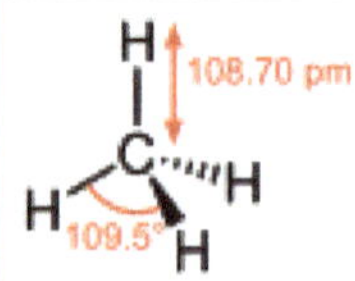	56
Propene	C_3H_6		44
Benzene	C_6H_6	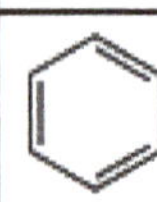	43
Acetylene	C_2H_2	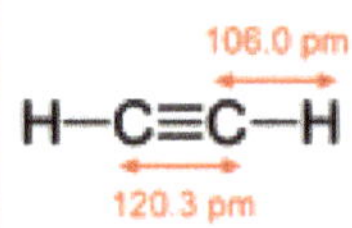	25
Cyclopentadiene	C_5H_6		18
Table 1. Carbon acid acidities in pKa in DMSO [4].			

Acid-base equilibria in molten salts. When a solid salt melts, it forms a solution of the cations and anions. For example, KOH melts at temperatures above 400 °C and dissociates into K^+ and OH^- ions which can act as a solvent for chemical reactions. Because of the autodissociation of the OH^- solvent, water is always present in a molten KOH flux, according to the acid-base equilibrium:

$$2\ OH^- \rightleftharpoons H_2O + O^{2-}$$

It follows that in this very basic solvent, water (the conjugate acid of the solvent) is the strongest acid that can exist. The conjugate base of the solvent, O^{2-}, is the strongest base. This autodissociation equilibrium allows for the acidity of a flux to be easily tuned through the addition or boiling off of water. A "wet" flux is more acidic, and can dissolve metal oxides that contain the basic O^{2-} anion. Conversely a "dry" flux is more basic and will cause oxides to precipitate. Molten hydroxide fluxes can thus be used in the synthesis of oxide crystals, such as the perovskite superconductor ($K_{1-X}Ba_XBiO_3$).[5]. Eutectic mixtures of NaOH and KOH are relatively low melting (≈ 200 °C) and can be used as solvents for crystallizing a variety of basic oxides.

Lewis Acids and Bases

The Lewis classification of acids and bases is broader than the Brønsted-Lowry definition, and encompasses many more substances. Whereas the Brønsted-Lowry and the Arrhenius classifications are based on transfer of protons, Lewis acidity and basicity are based on the sharing of an electron pair. **Lewis acids** can accept an electron pair, while **Lewis bases** can donate an electron pair. This definition encompasses the Brønsted-Lowry definition, in that H^+ is an electron pair acceptor (when interacting with a base), and a base is an electron pair donor in its interaction with H^+. This is illustrated below for the protonation of ammonia.

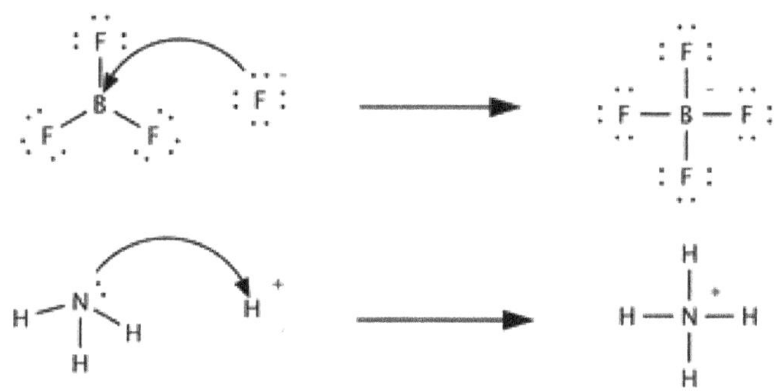

Boron trifluoride, BF_3 acts as a Lewis acid when it combines with a basic ion or molecule that can donate an electron pair. Such a reaction is shown below.

$BF_3 + F^- \rightleftharpoons BF_4^-$

Here, the acid is BF_3 and the base is F^-. This acid-base reaction allows boron (which is electron-deficient in BF_3) to complete its octet. Similarly, $AlCl_3$ is a Lewis acid that can react with Cl^- (a Lewis base) to make the Lewis "salt" $AlCl_4^-$. Note that in water Cl^- is a spectator ion (a weaker base than the solvent) in Brønsted acid-base reactions.

Additional examples of Lewis acid base reactions. In each, try to identify the acid, the base, and the salt, based on the concept that the base is the molecule or ion that donates an electron pair. In cases where you are not sure, it may help to draw the VSEPR structures of the molecules:

$I_2 + I^- \rightleftharpoons I_3^-$

$AuCl_3 + Cl^- \rightleftharpoons [AuCl_4]^-$

$Fe^{3+} + 6\ H_2O \rightleftharpoons [Fe(H_2O)_6]^{3+}$

$TiF_4 + 2\ F^- \rightleftharpoons [TiF_6]^{2-}$

$SF_4 + SbF_5 \rightleftharpoons$ "$SSbF_9$"

Lewis acidity is the basis for coordination chemistry. This is because coordination chemistry involves **metal ions** that are Lewis acids, which bond to ligands that are Lewis bases.**Determining the strength of metal ion Lewis acids**. There are three determining factors in the Lewis acid strength of a metal ion:1. The higher positive charge on the metal, the more acidic it is. For example, Al^{3+} and Fe^{3+} are good Lewis acids and their salts make acidic solutions in water, but K^+ and Na^+ are not.2. The smaller the atomic radius of the metal ion, the more acidic it is. Going down the periodic table, the Lewis acidity of metal ions decreases (e.g., $Al^{3+} > Ga^{3+} > In^{3+}$) because the ionic radius increases.3. For transition metal ions, more electronegative metals tend to make stronger Lewis acids. The electronegativity has maxima at W and Au in the 5d series, so metal ions near in that part of the periodic table are good Lewis acids.Molecules with five coordinate geometries (e.g., PCl_5, AsF_5, SbF_5) are typically strong Lewis acids, because when accepting another pair of electrons from a base, they form an octahedral molecule or anion. Neither of the common five-coordinate geometries (trigonal bipyramidal or square pyramidal) is efficient in terms of packing. The Lewis acid-base reaction forms an additional bond with a relatively small energetic penalty of stretching the existing bonds:

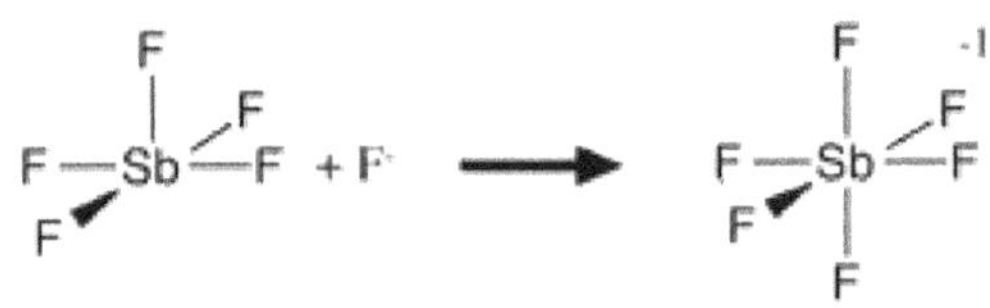

Because F^- is a good Lewis base, and also a small anion, it can form stable octahedral anions with both main group elements and transition metals. For this reason, TiO_2 and SiO_2 dissolve in HF (but are unreactive with aqueous HCl and other strong acids):

$$TiO_2 + 4\ HF + 2\ F^- \rightleftharpoons TiF_6^{2-} + 2\ H_2O$$

$$SiO_2 + 4\ HF + 2\ F^- \rightleftharpoons SiF_6^{2-} + 2\ H_2O$$

Electrolysis of HF: Henri Moissan isolating fluorine, which reacts with a sample in a test tube.

Lewis bases stabilize high oxidation states. An interesting example of using Lewis acid-base chemstry to drive reactions is the chemical synthesis of fluorine gas, which was devised by Karl O. Christe in 1986.[6] Christe at the time was organizing a symposium to commemorate the 100th anniversary of the isolation of elemental fluorine by Henri Moissan, which Moissan did in 1886 by electrolyzing a solution of anhydrous HF. 100 years later, there was still no direct (non-electrochemical) synthesis of F_2. Christe's reaction scheme followed two steps. The first was the known synthesis of K_2MnF_6 from $KMnO_4$:

$$4\ MnO_4^-(aq) + 10\ H_2O(l) + 24\ F^-(aq) \rightarrow 4\ MnF_6^{2-}(aq) + 3O_2(g) + 20\ OH^-(aq)$$

$$2\ K^+(aq) + MnF_6^{2-}(aq) \rightarrow K_2MnF_6(s)$$

The second step involved reacting K_2MnF_6 with the powerful Lewis acid SbF_5, to make metastable MnF_4, which decomposes spontaneously to MnF_3 and fluorine gas:

$$K_2MnF_6(s) + 2\ SbF_5(l) \rightarrow 2\ KSbF_6(s) + \text{“}MnF_4\text{”}(s)$$

$$\text{“}MnF_4\text{”}(s) \rightarrow MnF_3(s) + 1/2\ F_2(g)$$

This reaction teaches us something interesting and important about the connection between acid-base and redox chemistry. Acids tend to stabilize low oxidation states, and bases stabilize high oxidation states. Mn is stable in the +4 oxidation state in K_2MnF_6, where it is surrounded by six basic F^- anions. However, the highest stable neutral fluoride of Mn is MnF_3, and MnF_4 (transiently formed from K_2MnF_6) spontaneously decomposes to generate fluorine.

Oxide is a better base than fluoride. Interestingly, Mn can lose all its valence electrons to form Mn^{7+} in the permanganate ion, MnO_4^-. Here the 7+ oxidation state is stabilized electrostatically by coordination to four O^{2-} ions, and by the overall -1 charge on the MnO_4^- anion. Because of its 2- charge, O^{2-} is a stronger base and a better ion for stabilizing high oxidation states than F^-. This is a general trend among transition metals: the highest oxidation state is usually reached in the oxide, not in the fluoride, despite the fact that F is a more electronegative element than O. For example, Cr^{6+} is stable in the CrO_4^{2-} and $Cr_2O_7^{2-}$ anions, but not in any neutral fluoride or fluoroanion. The +8 oxidation state occurs in RuO_4 and OsO_4, but not in any fluoride of Ru or Os.

HARD AND SOFT ACIDS AND BASES

Lewis acids and bases can be classified by designating them as hard or soft.

Hard Acids/Bases:

"Hard" acids and bases have a high charge (positive for acids, negative for bases) to ionic radius ratio along with higher oxidation states. Hard acids interact with bases primarily through ionic bonding, and are not very polarizable. Hard acids generally have high charge densities. Thus metal ions with high positive charges and smaller ionic sizes tend to be hard acids. Early transition metal ions in the 3d series tend to be hard Lewis acids. Hard bases are typically small anions and neutral molecules. Some examples of hard acids and bases include: H^+, O^{2-}, OH^-, F^-, Fe^{3+}, and Al^{3+}.

Soft Acids/Bases:

"Soft" acids or bases have a low charge to radius ratio, with low oxidation states. They are normally larger ions that are polarizable. In contrast to the primarily ionic interaction of hard acids and bases, soft acids and bases have a significant covalent bonding component to their interaction. For example, I^- and S^{2-} are soft bases and low charge density transition metals, such as Ag^+, are considered soft acids. Soft acids often include transition metals in the second and third row of the periodic table that have a +1 or +2 charge, as well as late transition metals (especially those in the 4d and 5d series) with filled or almost completely filled *d* orbitals.

Acids and bases are not strictly hard or soft, with many ions and compounds being classified as intermediate. For example, trimethylborane, Fe^{2+}, and Pb^{2+} cations are intermediate acids, and pyridine and aniline are examples of intermediate bases. An element can also change its hard/soft character depending on its oxidation state. The most extreme example is hydrogen, where H^+ is a hard acid and H^- is a soft base. Ni^{3+} (as in the layered compound NiOOH) is a hard acid, but Ni^0 (as in $Ni(CO)_4$) is a soft acid. The figures below show hard/soft trends for acids (left) and bases (right) in the periodic table. For bases, the major hard/soft discontinuity is between the 2nd row (N,O,F) and the rows below.

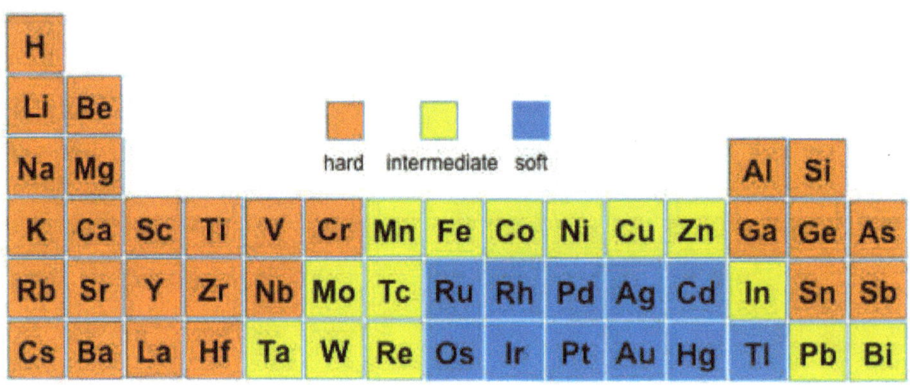

Hard-soft trends for acids.

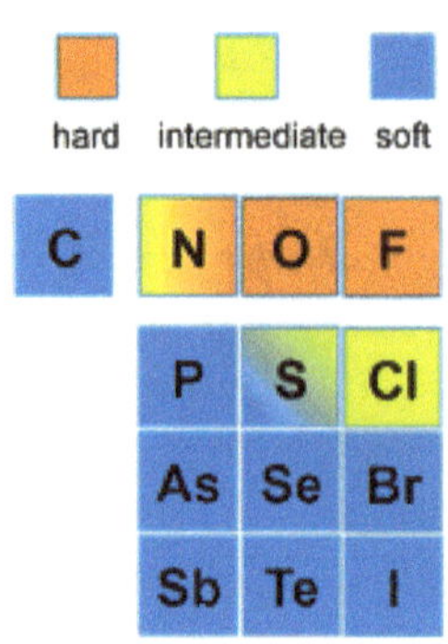

Hard-soft trends for bases.

Like binds with Like

Hard acids interact more strongly with hard bases than they do with soft bases, and soft acids interact more strongly with soft bases than hard bases. Thus, the most stable complexes are those with hard-hard and soft-soft interactions. This tendency is illustrated in the table below, which shows the trend in formation constants for hard and soft acids. Hard acids bind halides in the order $F^- > Cl^- > Br^- > I^-$, whereas soft acids follow the opposite trend.

Log K_1	fluoride	chloride	bromide	iodide	acid classification
Fe^{3+}	6.0	1.4	0.5	-	Hard
Pb^{2+}	1.3	0.9	1.1	1.3	Intermediate
Ag^+	0.4	3.3	4.7	6.6	Soft
Hg^{2+}	1.0	6.7	8.9	12.9	Soft

The softest metal ion in the periodic table is Au^+(aq). It forms stable complexes with soft bases such as phosphines and CN^-, but not with hard bases such as O^{2-} or F^-. The affinity of Au^+ for the soft base CN^- is high, and the resulting $[Au(CN)_2]^-$complex is so stable that gold (which is normally very difficult to oxidize) can be oxidized by oxygen in the air:

$$4\ Au(s) + 8\ CN^-(aq) + O_2(g) + 2\ H_2O = 4\ [Au(CN)_2]^-(aq) + 4\ OH^-$$

This reaction is used in gold mining to separate small flakes of Au from large volumes of sand and other oxides. Ag is similarly dissolved by air oxidation in cyanide solutions. The precious metals are then isolated from the solution using chemical reducing agents or by electroplating. The use of cyanide ion on a large scale in mining, however, creates a serious environmental hazard. Cyanide, which is highly toxic, is gradually oxidized by air to the less toxic cyanate (OCN^-) ion. On the laboratory scale, cyanide plating solutions are typically disposed of by using bleach to oxidize CN^- to OCN^-, and the metal is recovered as an insoluble chloride salt.

The Au^{3+} ion, because of its higher positive charge, is a harder acid than Au^+ and can form complexes with harder bases such as H_2O and amines. In keeping with the "like binds like" principle, the compound AuI (soft-soft) is stable, but AuI_3 (hard-soft) is unknown. Conversely, AuF has never been isolated but AuF_3 (hard-hard) is stable.

Discussion questions

- Discuss periodic trends in the Lewis acidity of metal ions.
- Explain what we mean by hard and soft acids and bases, using specific examples.

PROBLEMS

1. For each pair of compounds below, indicate which one is a stronger acid and explain your choice:

(a) $[Fe(H_2O)_6]^{3+}$ or $[Fe(H_2O)_6]^{2+}$

(b) H_3BO_3 or H_3PO_3

(c) $[Ga(H_2O)_6]^{3+}$ or $[Al(H_2O)_6]^{3+}$

(d) HIO_3 or HIO_4

(e) H_3PO_4 or H_2SeO_4

2. The Mg^{2+} and Cu^{2+} cations have similar ionic radii, but the acidities of aqueous nitrate solutions containing the two ions are different. Explain which ion is more acidic and why.

3. Cobalt reacts with excess fluorine to make CoF_3, but higher oxidation states of cobalt do not exist in binary fluorides. However, Co(IV) is stable in oxides such as CoO_2 and $BaCoO_3$. Explain why these oxides can stabilize a higher oxidation state of cobalt than fluorides.

4. (a) When solutions of the hydrosulfide (SH^-) salts of As^{3+}, Sb^{3+}, and Sn^{4+} are reacted with aqueous solutions of ammonium hydrosulfide ($NH_4^+SH^-$), the sulfide salts As_2S_3, Sb_2S_3, and SnS_2 precipitate. When excess aqueous Na_2S is added, however, these sulfides re-dissolve to form soluble anionic complexes. In contrast, solutions of Cu^{2+}, Pb^{2+}, Hg^{2+}, Bi^{3+} and Cd^{2+}precipitate as solid sulfides but do not re-dissolve in solutions that contain excess sulfide. We may consider the first group of ions to be amphoteric for soft acid-base reactions that involve SH^- instead of OH^-. The second group is more basic (less acidic) in the sense that it does not react with excess soft base. Use this information to locate the amphoteric diagonal in the periodic table for sulfides. Compare this with the diagonal that defines the amphoteric oxides, which dissolve in either acidic or basic solutions. Does your analysis agree with the description of S^{2-} as a softer base than O^{2-}?

(b) Au^+ is also precipitated by SH^- ions, and the sulfide Au_2S redissolves to form a soluble complex in excess SH^-. Does this fit the trend you discovered in part (a)? Does it make sense in terms of the electronegativity of Au? Explain.

REFERENCES

1. Hall NF, Conant JB (1927). "A Study of Superacid Solutions". *Journal of the American Chemical Society* **49** (12): 3062–70. doi:10.1021/ja01411a010.
2. George A. Olah, Schlosberg RH (1968). "Chemistry in Super Acids. I. Hydrogen Exchange and Polycondensation of Methane and Alkanes in $FSO_3H–SbF_5$ ("Magic Acid") Solution. Protonation of Alkanes and the Intermediacy of CH_5^+ and Related Hydrocarbon Ions. The High Chemical Reactivity of "Paraffins" in Ionic Solution Reactions". *Journal of the American Chemical Society* **90** (10): 2726–7. doi:10.1021/ja01012a066.
3. Herlem, Michel (1977). "Are reactions in superacid media due to protons or to powerful oxidising species such as SO_3or SbF_5?". *Pure & Applied Chemistry* **49**: 107–113. doi:10.1351/pac197749010107.

4. *Equilibrium acidities in dimethyl sulfoxide solution* Frederick G. Bordwell Acc. Chem. Res.; **1988**; 21(12) pp 456 - 463; DOI:10.1021/ar00156a004

5. R. J. Cava, et al. (1988). "Superconductivity near 30 K without copper: the $Ba_{0.6}K_{0.4}BiO_3$ perovskite". *Nature* **332**: 814–6. doi:10.1038/332814a0.

6. Christe, Karl O. (1986). "Chemical synthesis of elemental fluorine". Inorganic Chemistry 25 (21): 3721. doi:10.1021/ic00241a001.

CHAPTER 9

Ionic and Covalent Solids - Structures

Chapter Contents

- **Introduction**
- **Close-Packing and Interstitial Sites**
- **Structures Related to Nacl and Nias**
- **Tetrahedral Structures**
- **Layered Structures and Intercalation Reactions**
- **Bonding in Tis2, Mos2, and Pyrite Structures**
- **Spinel, Perovskite, and Rutile Structures**

Covalent bonds are characterized by the sharing of electrons between two or more atoms. These bonds mostly occur between nonmetals or between two of the same (or similar) elements.Two atoms with similar electronegativity will *not* exchange an electron from their outermost shell; the atoms instead *share* electrons so that their valence electron shell is filled.

INTRODUCTION

As we noted in our discussion of metal and alloy structures, there is an intimate connection between the structures and the physical properties of materials. As we "graduate" from simple metal structures based on sphere packings to more complex structures, we find that this is still true. In this chapter we will try to systematize the structures of inorganic solids - metal oxides, halides, sulfides, and related compounds - and develop some rules for which structures to expect based on electronegativity differences, hard-soft acid-base rules, and other periodic trends.

The morphology of twinned crystals of iron pyrite (FeS_2) is related to the underlying cubic symmetry of the unit cell. Like NaCl, the pyrite crystal structure can be thought of as a face-centered cubic array of anions (S_2^{2-}) with cations (Fe^{2+}) occupying all the octahedral holes.

Inorganic solids often have simple crystal structures, and some of these structures are adopted by large families of ionic or covalent compounds. Examples of the most common structures include NaCl, CsCl, NiAs, zincblende, wurtzite, fluorite, perovskite, rutile, and spinel. We will develop these structures systematically from the close packed and non-close packed lattices shown below. Some layered structures, such as $CdCl_2$ and CdI_2, can be thought of as relatives of simple ionic lattices with some atoms "missing."

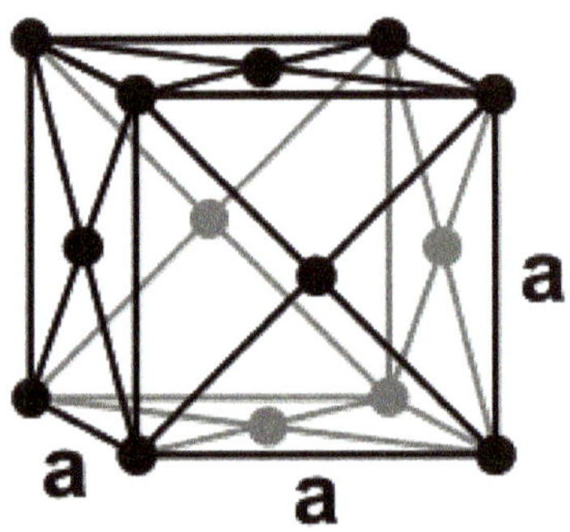

Face-centered cubic (fcc) or cubic close-packed (ccp)

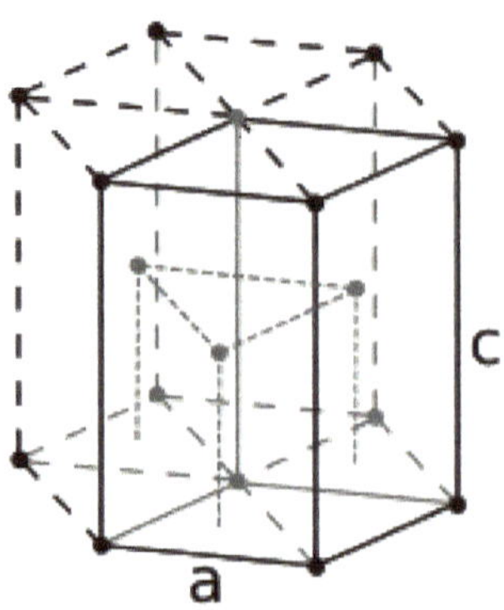

Hexagonal close-packed (hcp)

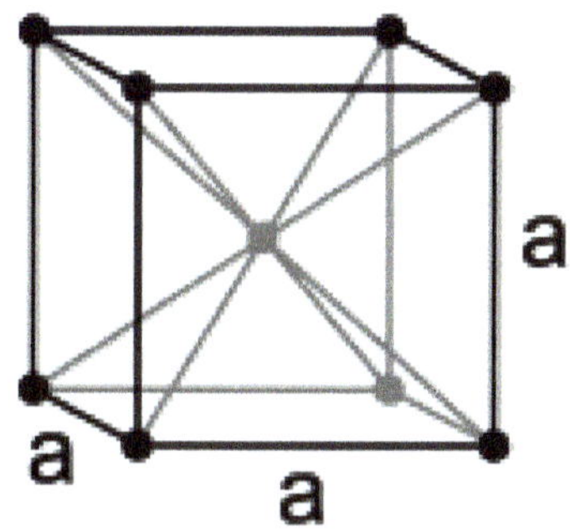

Body-centered cubic

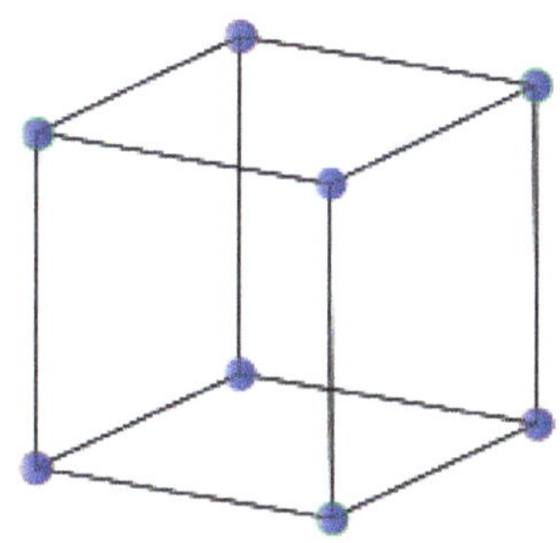

Simple cubic

CLOSE-PACKING AND INTERSTITIAL SITES

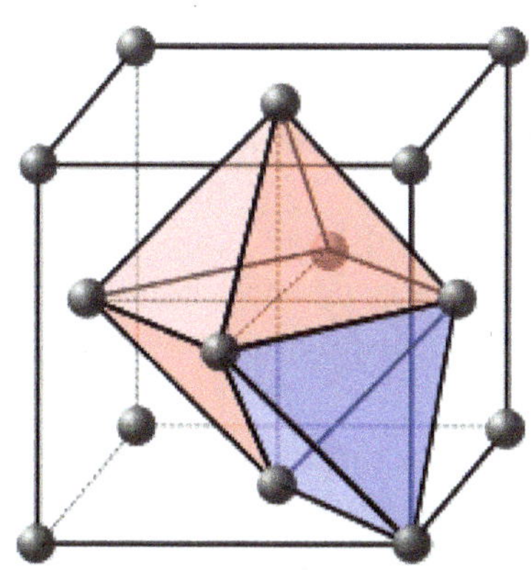

One octahedral and one tetrahedral site in a face-centered cubic unit cell. Each cell contains four packing atoms (gray), four octahedral sites (pink), and eight tetrahedral sites (blue).

Many common inorganic crystals have structures that are related to cubic close packed (face-centered cubic) or hexagonal close packed sphere packings. These packing lattices contain two types of sites or "holes" that the interstitial atoms fill, and the coordination geometry of these sites is either **tetrahedral** or **octahedral**. An interstitial atom filling a tetrahedral hole is coordinated to four packing atoms, and an atom filling an octahedral hole is coordinated to six packing atoms. In both the hexagonal close packed and cubic close packed lattices, there is one octahedral hole and two tetrahedral holes per packing atom.

We might expect that anions, which are often larger than cations, would be better suited to the positions of packing atoms. While this is often true, there are many examples of structures in which cations are the packing atoms, and others in which the distinction is arbitrary. The NaCl structure is a good example of the latter.

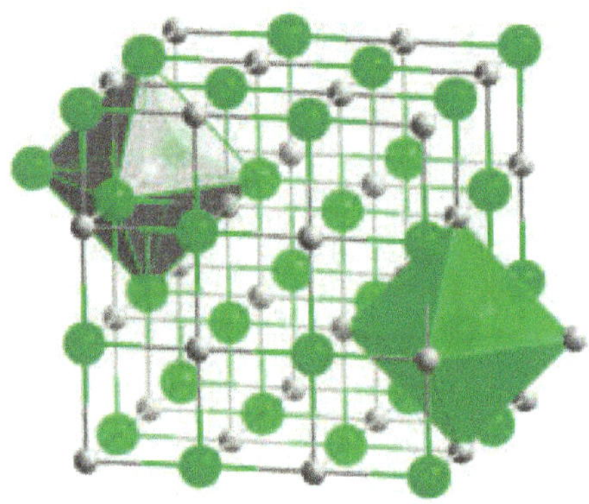

Crystal structure of NaCl. Both the Na^+ and Cl^- ions are octahedrally coordinated.

In the NaCl structure, shown on the right, the green spheres are the Cl^- ions and the gray spheres are the Na^+ ions. The octahedral holes in a face-centered cubic lattice can be found at fractional coordinates (1/2 1/2 1/2), (1/2 0 0), (0 1/2 0), and (0 0 1/2). There are four of these holes per cell, and they are filled by the chloride ions. The packing atoms (Na^+) have coordinates (0 0 0), (0 1/2 1/2), (1/2 1/2 0), and (1/2 0 1/2). Note that each of the Na^+ positions is related to a Cl^-position by a translation of (1/2 0 0). Another way of stating this is that the structure consists of **two interpenetrating fcc lattices**, which are related to each other by a translation of half the unit cell along any of the three Cartesian axes. We could have equivalently placed the Cl ions at the fcc lattice points and the Na ions in the octahedral holes by simply translating the origin of the unit cell by (1/2 0 0). Thus the distinction between packing and interstitial atoms in this case is arbitrary.

NaCl is interesting in that it is a three-dimensional checkerboard, and thus there are no NaCl "molecules" that exist in the structure. When this structure was originally solved (in 1913 by using X-ray diffraction) by W. L. Bragg, his interpretation met resistance by chemists who thought that precise integer stoichiometries were a consequence of the valency of atoms in molecules. The German chemist P. Pfeiffer noted in 1915 that 'the ordinary notion of valency didn't seem to apply', and fourteen years later, the influential chemist

H. E. Armstrong still found Bragg's proposed structure of sodium chloride 'more than repugnant to the common sense, not chemical cricket'! Nevertheless, Bragg and his father, W. H. Bragg, persevered and used the then-new technique of X-ray diffraction to determine the structures of a number of other compounds, including diamond, zincblende, calcium fluoride, and other alkali halides. These experiments gave chemists their first real look at the atomic structure of solids, and laid the groundwork for X-ray diffraction experiments that later elucidated the structures of DNA, proteins, and many other compounds. For their work on X-ray diffraction the Braggs received the Nobel prize in Physics in 1915.

Since each type of atom in the NaCl structure forms a face-centered cubic lattice, there are four Na and four Cl atoms per NaCl unit cell. It is because of this ratio that NaCl has a 1:1 stoichiometry. The shaded green and gray bipyramidal structures in the NaCl lattice show that the Na^+ ions are coordinated to six Cl^- ions, and vice versa. The NaCl structure can be alternatively drawn as a stacking of close-packed layer planes, AcBaCbAcBa... along the body diagonal of the unit cell. Here the uppercase letters represent the packing atoms, and the lower case letters are the interstitial atoms. This layered packing is illustrated below:

NaCl structure

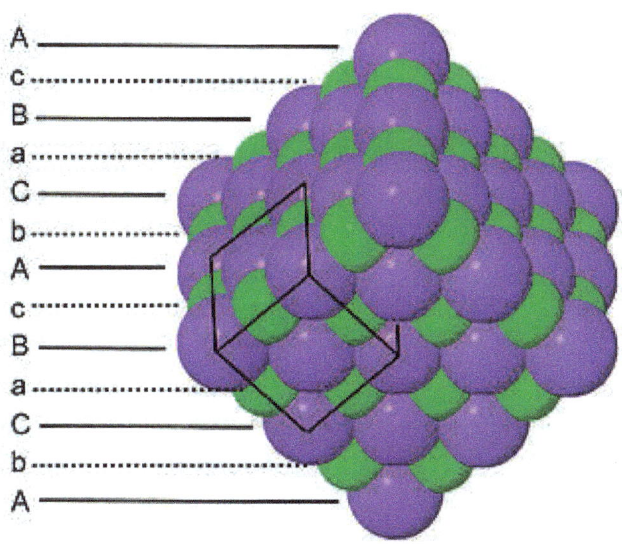

------------ A

- - -c- - - -

------------ B

- - -a- - - -

------------ C

- - -b- - - -

------------ A

- - -c- - - -

------------ B

- - -a- - - -

------------ C

- - -b- - - -

------------ A

Note that both the packing atoms and interstitials are stacked in the sequence A-B-C-A-B-C..., in keeping with the fact that each forms a cubic close-packed lattice.

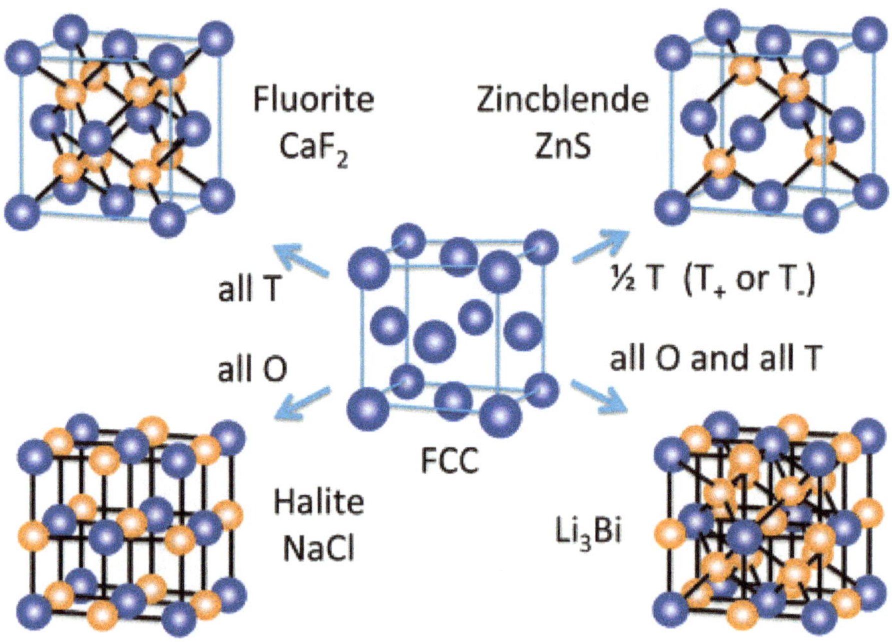

The NaCl structure is fairly common among ionic compounds:

- Alkali Halides (except CsCl, CsBr, and CsI)
- Transition Metal Monoxides (TiO, VO,..., NiO)
- Alkali Earth Oxides and Sulfides (MgO, CaO, BaS... except BeO and MgTe)
- Carbides and Nitrides (TiC, TiN, ZrC, NbC) -these are very stable refractory, interstitial alloys (metallic)

A number of other inorganic crystal structures are formed (at least conceptually) by filling octahedral and/or tetrahedral holes in close-packed lattices. The figure at the right shows some of the most common structures (fluorite, halite, zincblende) as well as a rather rare one (Li_3Bi) that derive from the fcc lattice. From the hcp lattice, we can make the NiAs and wurzite structures, which are the hexagonal relatives of NaCl and zincblende, respectively.

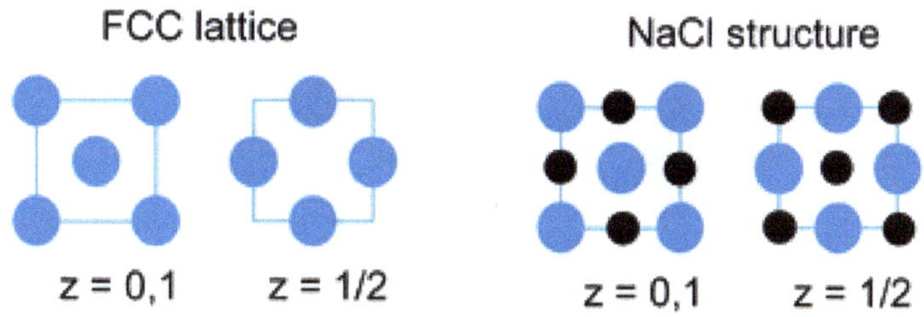

An alternative and very convenient way to represent inorganic crystal structures (especially complex structures such as Li_3Bi) is to draw the unit cell in slices along one of the unit cell axes. This kind of representation is shown at the left for the fcc lattice and the NaCl structure. Since all atoms in these structures have z-coordinates of either 0 or 1/2, only those sections need to be drawn in order to describe the contents of the unit cell. It is a useful exercise to draw some of the fcc compound structures (above right) in sections.

STRUCTURES RELATED TO NACL AND NIAS

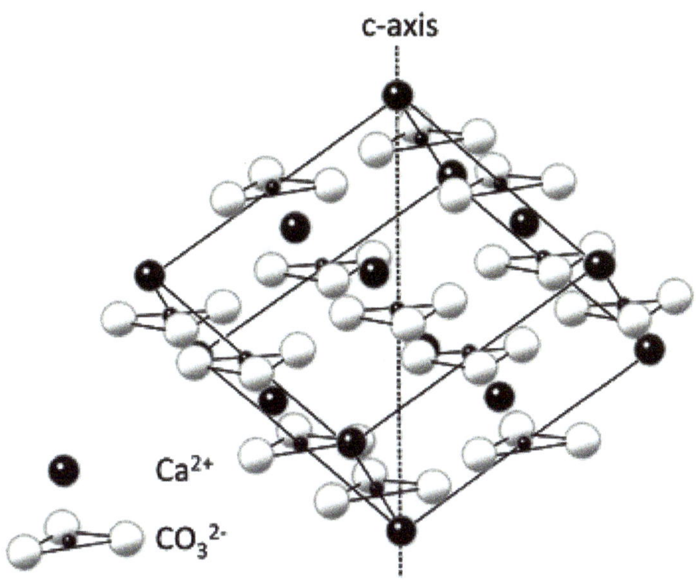

The rhombohedral unit cell of the calcite crystal structure. The hexagonal c-axis is shown.

There are a number of compounds that have structures similar to that of NaCl, but have a lower symmetry (usually imposed by the geometry of the anion) than NaCl itself. These compounds include:

FeS_2 (pyrite, “fools gold”): S_2^{2-} (disulfide) and Fe^{2+}

CaC_2 (a salt-like carbide): Ca^{2+} and linear C_2^{2-} anions

$CaCO_3$ (calcite, limestone, marble): Ca^{2+} and triangular CO_3^{2-}.

Calcite crystals are birefringent, meaning that their refractive indices are different along the two principal crystal directions. This gives rise to the phenomenon of double refraction.

The **calcite** ($CaCO_3$) crystal structure is shown at the right. Triangular CO_3^{2-} ions fill octahedral holes between the Ca^{2+}ions (black spheres) in a distorted NaCl lattice. As in NaCl, each ion is coordinated by six of the other kind. From this image we can see why the $CaCO_3$ structure has a lower symmetry than that of NaCl. The fourfold rotation symmetry of the NaCl unit cell is lost when the spherical Cl^- ions are replaced by triangular CO_3^{2-} ions. Because of this symmetry lowering, transparent crystals calcite are birefringent, as illustrated at the left.

NiAs structure. The NaCl structure can be described a face-centered cubic lattice with all of the octahedral holes filled. What if we start with a hexagonal-close packed lattice rather than a face-centered cubic lattice?

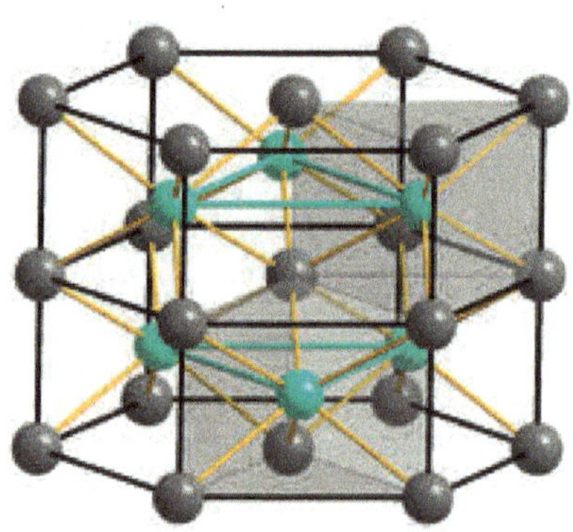

Nickel arsenide crystal structure. The Ni_6As trigonal prisms are shaded gray. One octahedron of six As atoms surrounding a Ni atom is shown in the center of the figure.

This is the structure adopted by **NiAs** and many other transition metal sulfides, phosphides, and arsenides. The cations are shown in gray while the anions are light blue in the figure at the right. The cations are in octahedral coordination, so each cation is coordinated to six anions. The anions are also coordinated to six cations, but they occupy trigonal prismatic sites. In terms of layer stacking, the NiAs structure is AcBcAcBc..., where the A and B sites (the hcp lattice) are occupied by the As atoms, and the c sites, which are eclipsed along the layer stacking axis, are occupied by Ni. Unlike the NaCl structure, where the anion and cation sites are interchangeable, NiAs has unique anion and cation sites. The layer stacking sequence for NiAs is shown below:

------------ A

- - -c- - - -

------------ B

- - -c- - - -

------------ A

- - -c- - - -

------------ B

- - -c- - - -

The NiAs structure cannot be adopted by ionic compounds because of the eclipsing cations, because the cation-cation repulsions would be internally destabilizing for an ionic compound. This structure is mainly adopted by covalent and polar covalent MX compounds, typically with "soft" X anions (S, Se, P, As,....) and low-valent transition metal cations. For example, some compounds with the NiAs structure are: MS, MSe, MTe (M=Ti, V, Fe, Co, Ni). Often these are nonstoichiometric or complex stoichiometries with ordered vacancies (Cr_7S_8, Fe_7S_8).

TETRAHEDRAL STRUCTURES

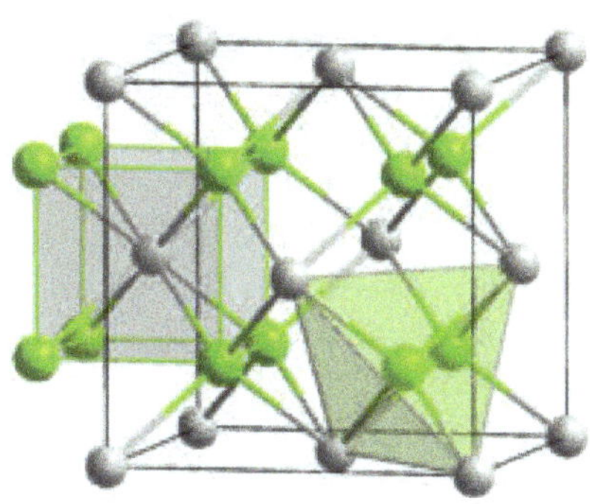

The fluorite (CaF_2) crystal structure showing the coordination environments of the Ca and F atoms.

In ccp and hcp lattices, there are two tetrahedral holes per packing atom. A stoichiometry of either M_2X or MX_2 gives a structure that fills all tetrahedral sites, while an MX structure fills only half of the sites. An example of an MX_2 structure is **fluorite**, CaF_2, whose structure is shown in the figure at the left. The packing atom in fluorite is Ca^{2+} and the structure is composed of three interpenetrating fcc lattices. It should be noted that the Ca^{2+} ion (gray spheres) as a packing atom defies our "rule" that anions are larger than cations and therefore must be the packing atoms. The fluorite structure is common for ionic MX_2 (MgF_2, ZrO_2, etc.) and M_2X compounds (Li_2O). In contrast, the hcp relative of the fluorite structure is quite rare because of unfavorable close contacts between like-charged ions.

In terms of geometry, Ca^{2+} is in cubic coordination with eight F^- neighbors, and the fluoride ions are tetrahedrally coordinated by four Ca^{2+} ions. The 8:4 coordination geometry is consistent with the 1:2 Ca:F stoichiometry; in all crystal structures the ratio of the coordination numbers is the inverse of the stoichiometric ratio. The three interpenetrating fcc lattices have Ca at 0,0,0 , 1/2,1/2,0 , etc....F at 1/4,1/4,1/4 , 3/4,3/4,1/4 , etc... and F at 3/4,3/4,3/4 , 1/4,1/4/3/4 , etc.

Looking more closely at the tetrahedral sites in fluorite, we see that they fall into two distinct groups: T_+ and T_-. If a tetrahedron is oriented with a vertex pointing upwards along the stacking axis, the site is T_+. Likewise, a tetrahedron with a vertex oriented downward is T_-. The alternation of T_+ and T_- sites allows for efficient packing of ions in the structure. The layer stacking sequence in this structure (including fluoride ions in the T_+ and T_- sites) is:

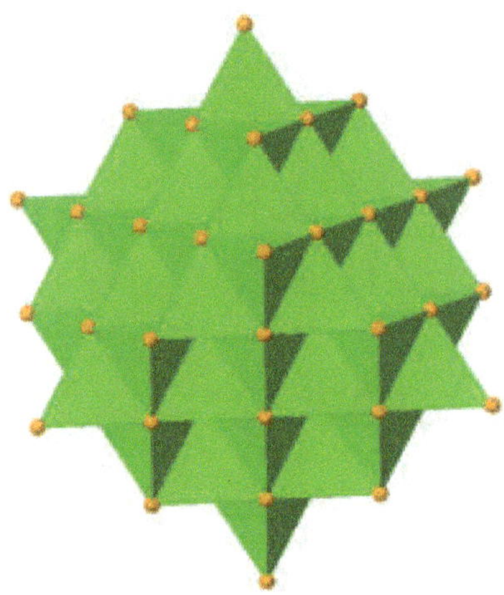

Polyhedral view of the fluorite crystal structure, showing T_+ and T_-Ca_4F tetrahedra. The Ca^{2+} ions are stacked ABCABC... along the body diagonal of the unit cell, which is the vertical direction in this image.

------------ A

- - -b- - - T_+

- - -a- - - T_-

------------ B

- - -c- - - T_+

- - -b- - - T_-

------------ C

- - -a- - - T_+

- - -c- - - T_-

------------ A

- - -b- - - T_+

- - -a- - - T_-

Tetrahedrally bonded compounds with a 1:1 stoichiometry (MX compounds) have only **half** of the tetrahedral sites (either the T_+ or T_- sites) filled. In this case, both the M and the X atoms are tetrahedrally coordinated. The **zincblende** and **wurtzite** structures are 1:1 tetrahedral structures based on fcc and hcp lattices, respectively. Both structures are favored by p-block compounds that follow the octet rule, and these compounds are usually semiconductors or insulators. The zincblende structure, shown below, can be thought of as two interpenetrating fcc lattices, one of anions and one of cations, offset from each other by a translation of 1/4 along the body diagonal of the unit cell. Examples of compounds with the zincblende structure include CuCl, CuI, ZnSe, HgS, BeS, CdTe, AlP, GaP, SnSb, CSi, and diamond. Additionally, the compound $CuInSe_2$ is zincblende in an ordered, doubled unit cell (the chalcopyrite structure). The solid solution compounds $CuIn_{1-x}Ga_xSe_2$ with this structure are among the most widely studied materials for use in efficient thin film photovoltaic cells. Using ZnS as a representative of zincblende, the coordination of both Zn and S atoms is tetrahedral. The layer sequence, which is AbBcCaAbBcC..., results in six-membered ZnS rings that have the same geometry as the "chair" version of cyclohexane. The chair conformation allows for a relatively long distance between opposite atoms in the ring and, as a result, it is more sterically favorable than the boat form. The sequence of close-packed layers in zincblende, filling only the T_+ sites and leaving the T_- sites empty, is shown below:

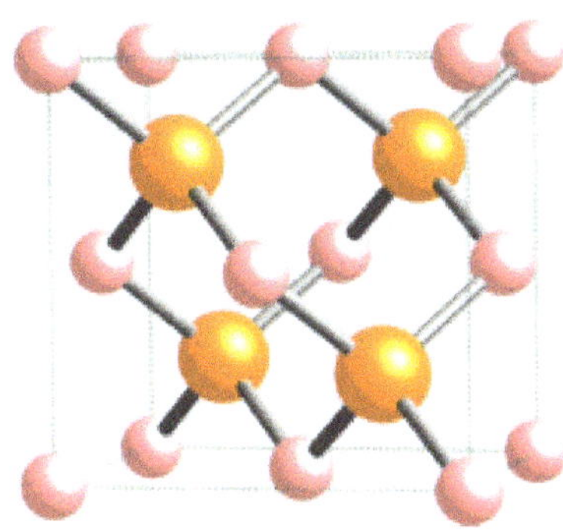

The zincblende unit cell

------------ A

- - -b- - - T_+

- - - - - - T_-

------------ B

- - -c- - - T_+

- - - - - - T_-

------------ C

- - -a- - - T_+

- - - - - - T_-

The wurtzite structure is a close relative of zinc blende, based on filling half the tetrahedral holes in the hcp lattice. Like zincblende, wurtzite contains planes of fused six-membered rings in the chair conformation. Unlike zincblende, however, the rings joining these planes contain six-membered "boat" rings. The boat aligns the anions so that they are directly above the cations in the structure, a less favorable situation sterically but a more favorable one in terms of electrostatics. As a result, the wurtzite structure tends to favor more polar or ionic compounds (e.g., ZnO, $NH_4^+F^-$) than the zincblende structure. As with zincblende, both ions are in tetrahedral (4:4) coordination and there are typically eight valence electrons in the MX compound. Examples of compounds with this structure include: BeO, ZnO, MnS, CdSe, MgTe, AlN, and NH_4F. The layered structure of wurtzite is AbBaAbB and the layer sequence with T_+ sites filled is illustrated below:

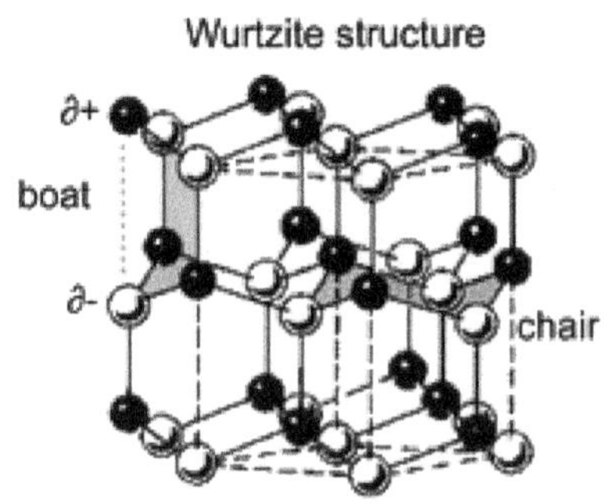

The chair and boat conformations of six-membered ZnS rings in the wurtzite structure.

------------ A

- - -b- - - T_+

- - - - - - T_-

------------ B

- - -a- - - T_+

- - - - - - T_-

------------ A

- - -b- - - T_+

- - - - - - T_-

------------ B

An interesting consequence of the layer stacking in the wurtzite structure is that the crystals are polar. When cleaved along the c-axis (the stacking axis), crystals of ZnO, ZnS, and GaN have one negatively charged face and an opposite positively charged face. An applied electric field interacts with the crystal dipole, resulting in compression or elongation of the lattice along this direction. For this reason crystals of compounds in the wurtzite structure are typically piezoelectric.

Some compounds are diamorphic and can have either the zincblende or wurtzite structure. Examples of these compounds that have intermediate polarities include CdS and ZnS. SiO_2 exists in polymorphs (crystobalite and tridymite) that resemble zincblende and wurtzite with O atoms midway between each of the Si atoms. The zincblende and wurtzite structures have efficient packing arrangements for tetrahedrally bonded networks and are commonly found in compounds that have tetrahedral bonding. Water, for example, has a tetrahedral hydrogen bonding network and is wurtzite-type. The undistorted wurtzite and zinc blende structures are typically found for AX compounds with eight valence electrons, which follow the octet rule. AX compounds with nine or ten electrons such as GaSe and GaAs crystallize in distorted variants of the wurtzite structure. In GaSe, the extra electrons form lone pairs and this creates layers in the structure, as can be seen in the figure below. To the right of GaSe, the structures of As, Sb, and SbAs show an ever further breakdown of the structure into layers as more valence electrons are added.

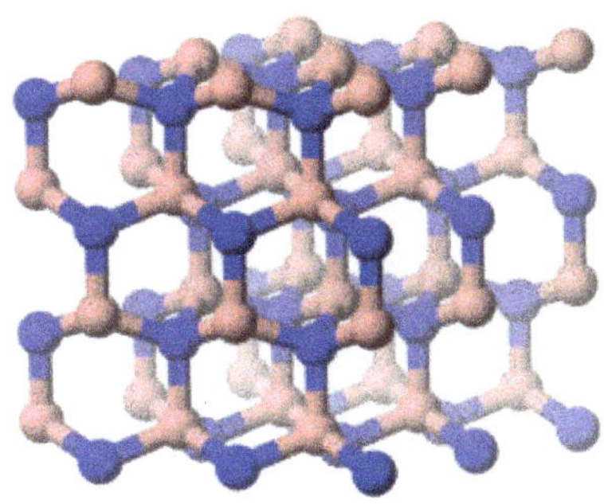

Wurtzite crystal structure

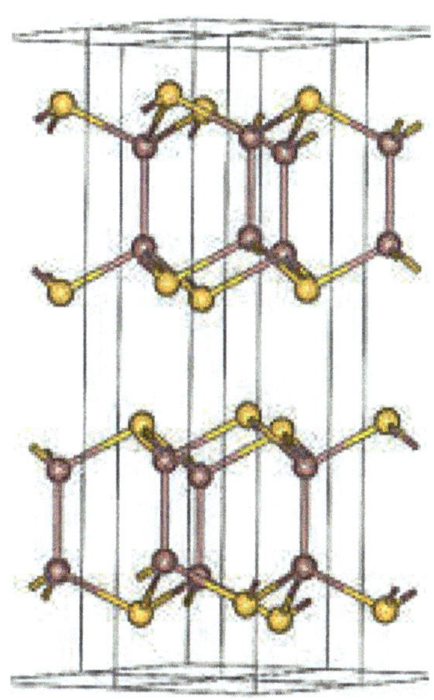

GaSe crystal structure

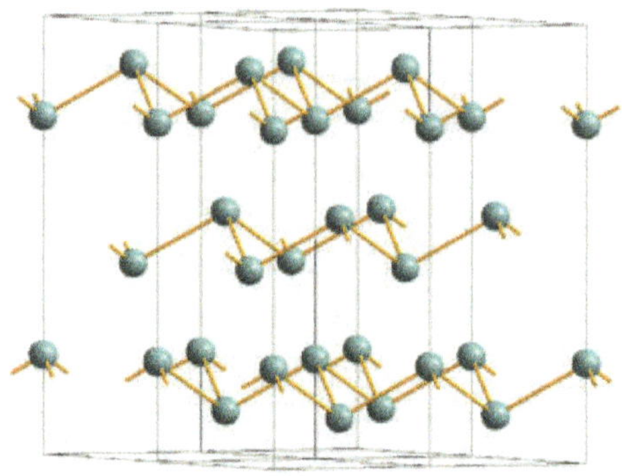

Crystal structure of Sb, SbAs, and gray As

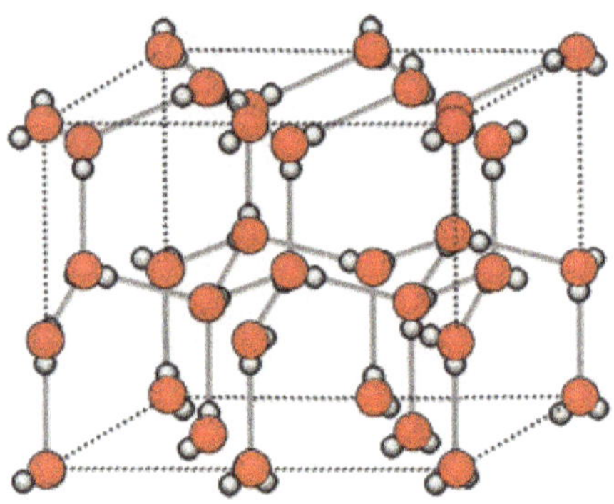

Crystal structure of hexagonal ice.

Hexagonal ice is the most stable polymorph of ice, which is obtained upon freezing at 1 atmosphere pressure. This polymorph (ice-I) has a hcp wurtzite-type structure. Looking at the structure shown at the right, we see that there are irregular arrangements of the O-H---O bonds. In the structure, hydrogen bonding enforces the tetrahedral coordination of each water molecule, resulting in a relatively open structure that is less dense than liquid water. For this reason, ice floats in water.

LAYERED STRUCTURES AND INTERCALATION REACTIONS

Layered structures can form when some fraction of the octahedral and/or tetrahedral sites are filled in the fcc and hcp lattices. For example, the $CdCl_2$ structure is formed by filling all the octahedral sites in alternate layers of the fcc lattice, and the CdI_2 structure is the relative of this structure in the hcp lattice.

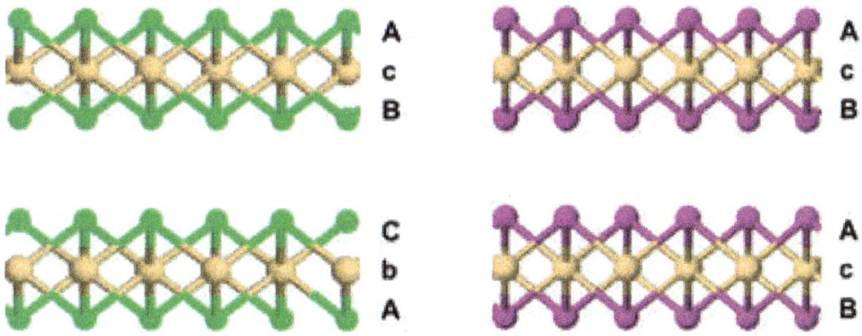

Comparison of the $CdCl_2$ (left) and CdI_2 (right) crystal structures.

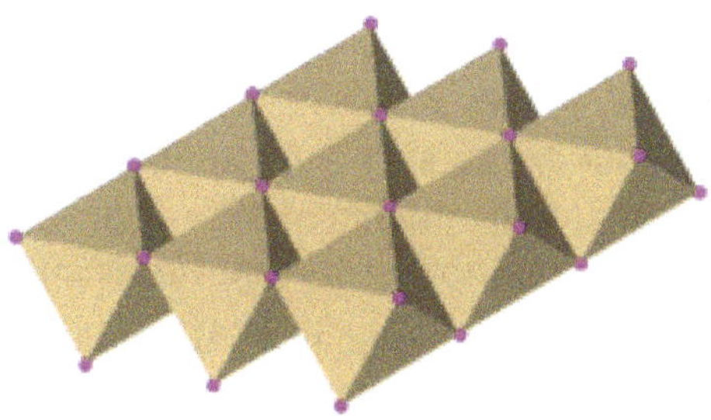

Polyhedral drawing of one layer of the $CdCl_2$ or CdI_2 structure showing edge-sharing MX_6 octahedra.

In the **$CdCl_2$structure**, the stacking sequence of anion layers is ABCABC...

In the **CdI_2structure**, the anion stacking sequence is ABAB..., and all the cations are eclipsed along the stacking axis.

These are examples of **6-3 structures**, because the cations are coordinated by an octahedron of six anions, and the anions are coordinated by three cations to make a trigonal pyramid (like NH_3). Another way to describe these structures is to say that the MX_6 octahedra each share six edges in the MX_2 sheets.

Because these structures place the packing atoms (the anions) in direct van der Waals contact, they are most stable for relatively **covalent** compounds. Otherwise, the electrostatic repulsion between contacting anions would destabilize the structure energetically. More ionic MX_2 compounds tend to adopt the fluorite (CaF_2) or rutile (TiO_2) structures, which are not layered.

Despite the fact that these two structure types are the same at the level of nearest and next-nearest neighbor ions, the CdI_2structure is much more common than the $CdCl_2$ structure.

$CdCl_2$ structure:

MCl_2 (M = Mg, Mn, Fe, Co, Ni, Zn, Cd)

$NiBr_2$, NiI_2, $ZnBr_2$, ZnI_2

CdI_2 structure:

MCl_2 (M = Ti, V)

MBr_2 (M = Mg, Fe, Co, Cd)

MI_2 (M = Mg, Ca, Ti, V, Mn, Fe, Co, Cd, Ge, Pb, Th)

$M(OH)_2$ (M = Mg, Ca, Mn, Fe, Co, Ni, Cd)

MS_2 (M = Ti, Zr, Sn, Ta, Pt)

MSe_2 (M = Ti, Zr, Sn, V, Pt)

MTe_2 (M = Ti, Co, Ni, Rh, Pd, Pt)

Physically, layered compounds are **soft** and **slippery**, because the layer planes slide past each other easily. For example, graphite, MoS_2, and talc (a silicate) are layered compounds that are used widely as **lubricants** and lubricant additives.

An important reaction of layered compounds is **intercalation**. In intercalation reactions, guest molecules and ions enter the galleries that separate the sheets, usually with expansion of the lattice along the stacking axis. This reaction is typically reversible if it does not perturb the bonding within the sheets. Often the **driving force for intercalation is a redox reaction**, i.e., electron transfer between the host and guest. For example, lithium metal reacts with TiS_2, MoS_2, and graphite to produce $LiTiS_2$, Li_xMoS_2 ($x < 1$), and LiC_6. In these compounds, lithium is ionized to Li^+ and the sheets are negatively charged. Oxidizing agents such as Br_2, $FeCl_3$, and AsF_5 also react with graphite. In the resulting intercalation compounds, the sheets are positively charged and the intercalated species are anionic.

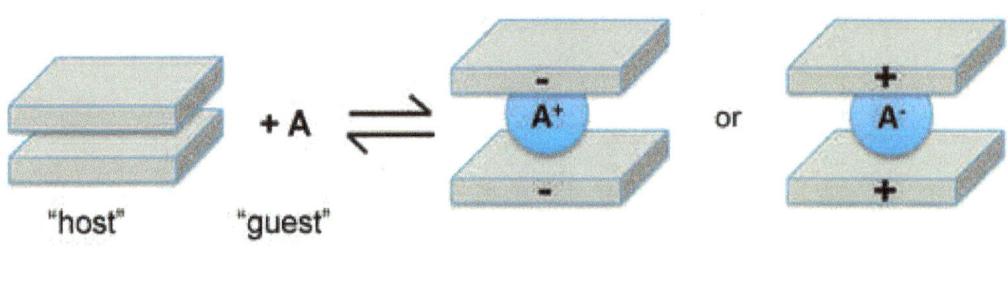

Oxidative or reductive intercalation involves the placement of anions or cations between sheets.

Intercalation reactions are especially important for electrochemical energy storage in **secondary batteries**, such as **lithium ion** batteries, **nickel-metal hydride** batteries, and **nickel-cadmium** batteries. The reversible nature of the intercalation reaction allows the electrodes to be charged and discharged up to several thousand times without losing their mechanical integrity. In lithium ion batteries, the negative electrode material is typically graphite, which is intercalated by lithium to make LiC_6. Several different oxides and phosphates containing redox active transition metal ions (Mn, Fe, Co, Ni) are used as the positive electrode materials.

Lithium ion batteries based on CoO_2 were first described in 1980[1] by John B. Goodenough›s research group at Oxford. In batteries based on CoO_2, which has the CdI_2 structure, the positive electrode half-reaction is:

$$LiCoO_2 \leftrightarrows \mathrm{Li}_{1-x}\mathrm{CoO}_2 + x\mathrm{Li}^+ + x\mathrm{e}^-$$

The negative electrode half reaction is:

$$x\mathrm{Li}^+ + x\mathrm{e}^- + x\mathrm{C}_6 \leftrightarrows x\mathrm{LiC}_6$$

The battery is fully charged when the positive electrode is in the CoO_2 form and the negative electrode is in the LiC_6 form. Discharge involves the motion of Li^+ ions through the electrolyte, forming Li_xCoO_2 and graphite at the two electrodes.

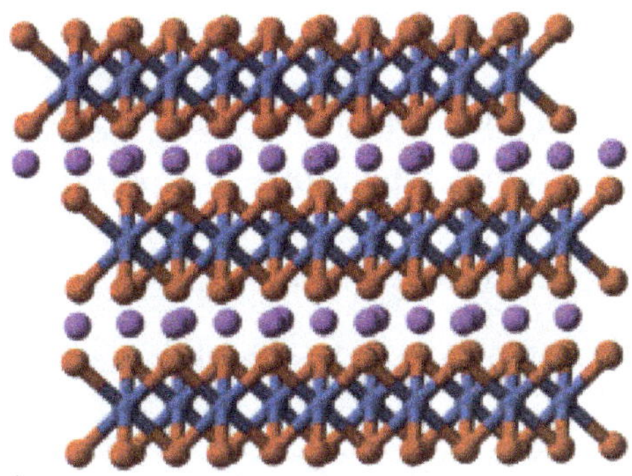

Crystal structure of $LiCoO_2$[2]

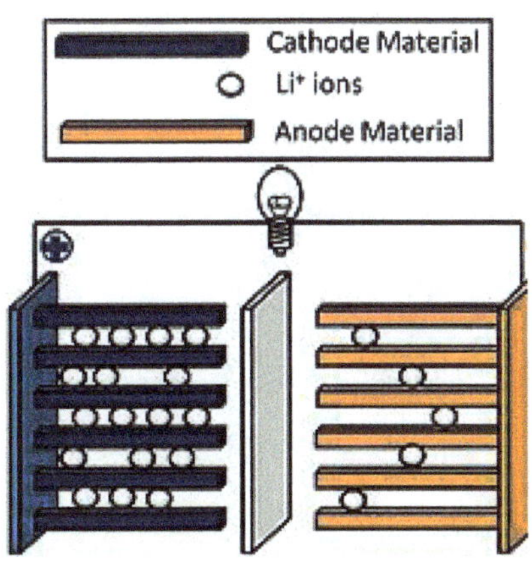

The lithium ion battery is a "rocking chair" battery, so named because charging and discharging involve moving Li^+ ions from one side to the other. CoO_2 is one example of a positive electrode material that has been used in lithium ion batteries. It has a high energy density, but batteries based on CoO_2 have poor thermal stability. Safer materials include lithium iron phosphate ($LiFePO_4$), and $LiMO_2$ (M = a mixture of Co, Mn, and Ni). These batteries are used widely in laptop computers, portable electronics, cellular telephones, cordless tools, and electric and hybrid vehicles.

A similar intercalation reaction occurs in nickel-cadmium batteries and nickel-metal hydride batteries, except in this case the reaction involves the movement of protons in and out of the $Ni(OH)_2$ lattice, which has the CdI_2 structure:

$$NiO(OH) + H_2O + e^- \rightarrow Ni(OH)_2 + OH^-$$

There are many layered compounds that cannot be intercalated by redox reactions, typically because some other stable product is formed. For example, the reaction of layered CdI_2 with Li produces LiI (NaCl structure) and Cd metal.

BONDING IN TIS_2, MOS_2, AND PYRITE STRUCTURES

Many layered dichalcogenides, such as TiS_2 and ZrS_2, have the CdI_2 structure. In these compounds, as we have noted above, the metal ions are octahedrally coordinated by S. Interestingly, the structures of MoS_2 and WS_2, while they are also layered, are different. In these cases, the metal is surrounded by a **trigonal prism** of sulfur atoms. NbS_2, TaS_2, $MoSe_2$, $MoTe_2$, WSe_2, and WTe_2 also have the trigonal prismatic molybdenite structure, which is shown below alongside a platy crystal of MoS_2.

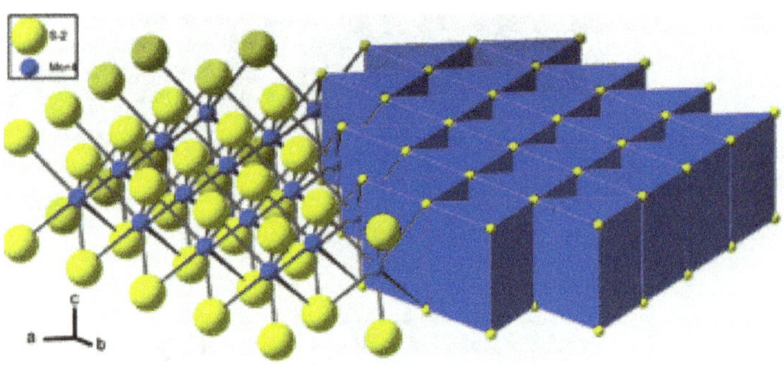

The coordination of the metal ions by a trigonal prism of chalcogenide ions is **sterically unfavorable** relative to octahedral coordination. There are close contacts between the chalcogenide ions, which are eclipsed in the stacking sequence **AbA/BaB/AbA/BaB**... (where "/" indicates the van der Waals gap between layers). What stabilizes this structure?

The molybdenite structure occurs most commonly in MX_2 compounds with a **d^1 or d^2 electron count**. The figure below compares the splitting of d-orbital energies in the octahedral and trigonal prismatic coordination environments:

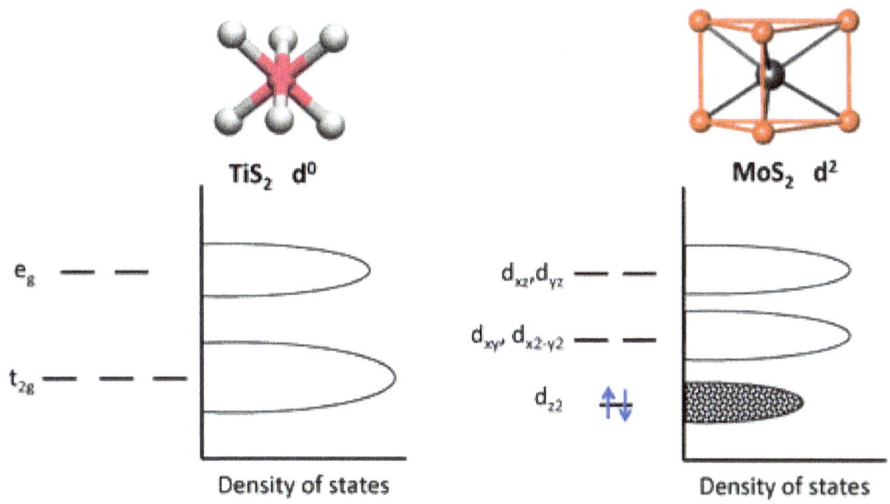

d-orbital splittings and energy bands in TiS_2 and MoS_2. MoS_2 is a semiconductor with a 1.3 eV gap between its filled and empty bands.

The trigonal prismatic structure is stabilized in MoS_2 by filling the lowest energy band, the d_{z2}. The d_{z2} orbital which points vertically through the triangular top and bottom faces of the trigonal prism, has the least interaction with the sulfide ligands and therefore the lowest energy. The d_{xz} and d_{yz} orbitals, which point at the ligands, have the highest energy. The d_{z2} orbital is lower in energy in this structure than the t_{2g} orbitals are in the octahedral structure of TiS_2.

PtS_2, like TiS_2, adopts the octahedral CdI_2structure. In this case, because Pt^{4+} has six d-electrons, the t_{2g} orbitals are filled. There is a large crystal field stabilization energy (which stabilizes the high oxidation state of Pt) because S^{2-} is a strong field ligand. Like MoS_2, PtS_2 is semiconducting because there is an energy gap between the filled t_{2g} and empty e_g bands.

Although early (TiS_2) and late (PtS_2) transition metal disulfides have layered structures, a number of MS_2 compounds in the middle of the transition series, such as MnS_2, FeS_2 and RuS_2, have three-dimensionally bonded structures. For example, FeS_2 has the **pyrite structure**, which is related to the NaCl structure. The reason is that FeS_2 is not $Fe^{4+}(S^{2-})_2$, but is actually $Fe^{2+}(S_2^{2-})$, where S_2^{2-} is the disulfide anion (which contains a single bond like the peroxide anion O_2^{2-}). S^{2-} is too strong a reducing agent to exist in the same compound with Fe^{4+}, which is a strong oxidizing agent. Because FeS_2 is actually $Fe^{2+}(S_2^{2-})$, it is a 1:1 compound and adopts a 1:1 structure.

Because it has an unfilled t_{2g} band, TiS_2 is relatively easy to reduce by intercalation with Li. For this reason, $LiTiS_2$ was one of the first intercalation compounds studied by

Stanley Whittingham, who developed the concept of the non-aqueous lithium ion battery in the early 1970›s.[3] Because it has a filled d_{z2} band, MoS_2 is harder to reduce, but it can be intercalated by reaction with the powerful reducing agent n-butyllithium to make Li_xMoS_2 ($x < 1$). Atoms in the van der Waals planes of these compounds are relatively unreactive, which gives MoS_2 its good oxidative stability and enables its application as a high temperature lubricant. Atoms at the edges of the crystals are however more reactive and in fact are catalytic. High surface area MoS_2, which has a high density of exposed edge planes, is used as a hydrodesulfurization catalyst and is also of increasing interest as an electrocatalyst for the reduction of water to hydrogen.

Layered metal dichalcogenides, including MoS_2, WS_2, and SnS_2, can form closed nanostructures that take the shape of multiwalled onions and multiwalled tubes. These materials were discovered by the group of Reshef Tenne in 1992, shortly after the discovery of carbon nanotubes. Since then nanotubes have been synthesized from many other materials, including vanadium and manganese oxides.

SPINEL, PEROVSKITE, AND RUTILE STRUCTURES

There are three more structures, which are derived from close-packed lattices, that are particularly important because of the material properties of their compounds. These are the **spinel** structure, on which ferrites and other magnetic oxides are based, the **perovskite** structure, which is adopted by ferroelectric and superconducting oxides, and the **rutile** structure, which is a common binary 6:3 structure adopted by oxides and fluorides.

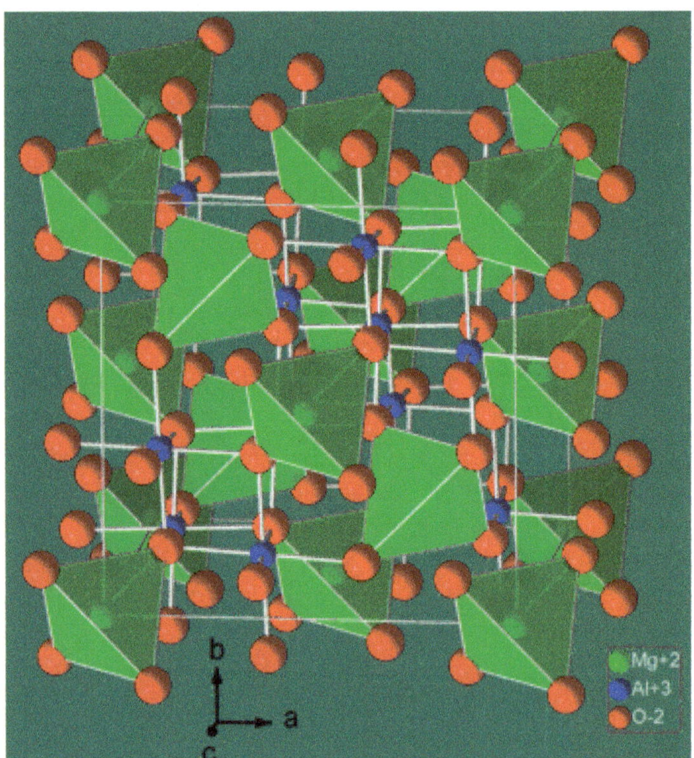

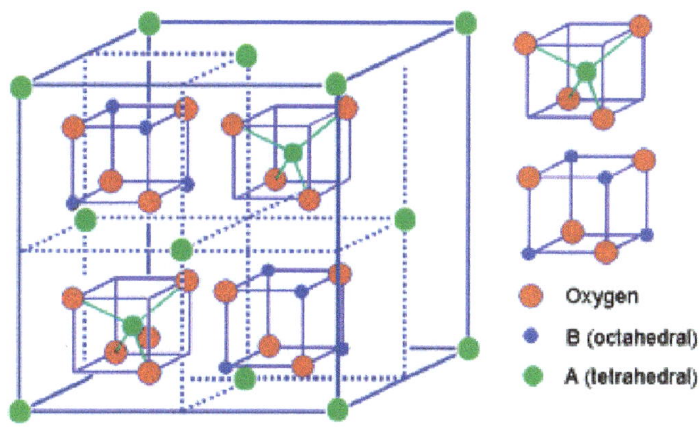

The **spinel structure** is formulated MM'_2X_4, where M and M' are tetrahedrally and octahedrally coordinated cations, respectively, and X is an anion (typically O or F). The structure is named after the mineral $MgAl_2O_4$, and oxide spinels have the general formula AB_2O_4.

In the **normal spinel** structure, there is a close-packed array of anions. The A-site cations fill 1/8 of the tetrahedral holes and the B-site cations fill 1/2 of the octahedral holes. A polyhedral view of the normal spinel unit cell is shown at the left, and a simplified view (with the contents of the back half of the cell removed for clarity) is shown above. Each unit cell contains eight formula units and has a composition $A_8B_{16}O_{32}$.

Inverse spinels have a closely related structure (with the same large unit cell) in which the A-site ions and half of the B-site ions switch places. Inverse spinels are thus formulated $B(AB)O_4$, where the AB ions in parentheses occupy octahedral sites, and the other B ions are on tetrahedral sites. There are also mixed spinels, which are intermediate between the **normal** and **inverse** spinel structure.

Some spinel and inverse spinel AB combinations are:

$A^{2+}B^{3+}$, e.g., $MgAl_2O_4$ (normal spinel)

$A^{4+}B^{2+}$, e.g., $Pb_3O_4 = Pb^{II}(Pb^{II}Pb^{IV})O_4$ (inverse spinel)

$A^{6+}B^{+}$, e.g., Na_2WO_4 (normal spinel)

Many magnetic oxides, such as Fe_3O_4 and $CoFe_2O_4$, are spinels.

Normal vs. inverse spinel structure. For transition metal oxide spinels, the choice of the normal vs. inverse spinel structure is driven primarily by the **crystal field stabilization energy** (CFSE) of ions in the tetrahedral and octahedral sites. For spinels that contain 3d elements such as Cr, Mn, Fe, Co, and Ni, the electron configuration is typically **high spin** because O^{2-} is a **weak field ligand**.

As an example, we can consider magnetite, Fe_3O_4. This compound contains one Fe^{2+} and two Fe^{3+} ions per formula unit, so we could formulate it as a normal spinel, $Fe^{2+}(Fe^{3+})_2O_4$, or as an inverse spinel, $Fe^{3+}(Fe^{2+}Fe^{3+})O_4$. Which one would have the lowest energy?

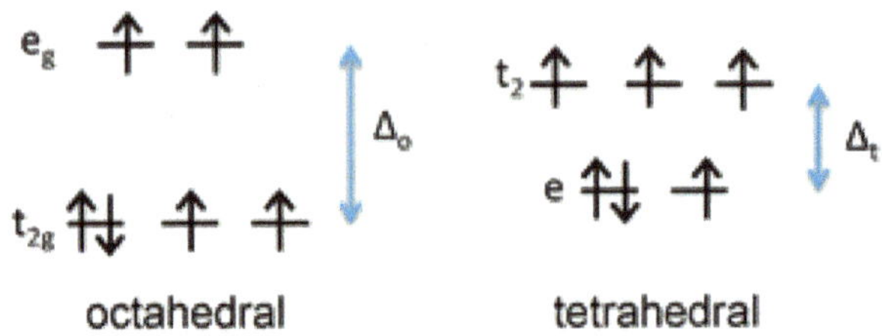

d-orbital energy diagram for Fe^{2+}

First we consider the crystal field energy of the $\mathbf{Fe^{2+}}$ ion, which is d^6. Comparing the tetrahedral and high spin octahedral diagrams, we find that the CFSE in an **octahedral** field of O^{2-} ions is $[(4)(2/5) - (2)(3/5)]\Delta_o - P = \mathbf{0.4\ \Delta_o - P}$. In the **tetrahedral** field, the CFSE is $[(3)(3/5) - (3)(2/5)]\Delta_t - P = \mathbf{0.6\ \Delta_t - P}$. Since Δ_o is about 2.25 times larger than Δ_t, the octahedral arrangement has a larger CFSE and is preferred for Fe^{2+}.

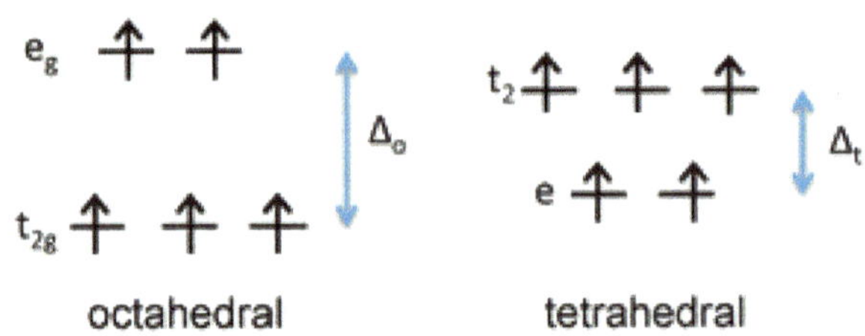

d-orbital energy diagram for Fe^{3+}

In contrast, it is easy to show that $\mathbf{Fe^{3+}}$, which is d^5, would have a CFSE of zero in either the octahedral or tetrahedral geometry. This means that Fe^{2+} has a preference for the octahedral site, but Fe^{3+} has no preference. Consequently, we place Fe^{2+} on octahedral sites and $\mathbf{Fe_3O_4}$ **is an inverse spinel**, $Fe^{3+}(Fe^{2+}Fe^{3+})O_4$.

Ferrites are compounds of general formula $\mathbf{M^{II}Fe_2O_4}$. We can see that magnetite is one example of a ferrite (with M = Fe). Other divalent metals (M = Mg, Mn, Co, Ni, Zn) also form ferrites. Ferrites can be normal or inverse spinels, or mixed spinels, depending on the CFSE of the M^{II} ion. Based on their CFSE, Fe^{2+}, Co^{2+}, and Ni^{2+} all have a strong preference for the octahedral site, so those compounds are all inverse spinels. $ZnFe_2O_4$ is a normal spinel because the small Zn^{2+} ion (d^{10}) fits more easily into the tetrahedral site than Fe^{3+} (d^5), and both ions have zero CFSE. $MgFe_2O_4$ and $MnFe_2O_4$, in which all ions have zero CFSE and no site preference, are mixed spinels. **Chromite** spinels, $\mathbf{M^{II}Cr_2O_4}$, are always **normal spinels** because the d^3 Cr^{3+} ion has a strong preference for the octahedral site.

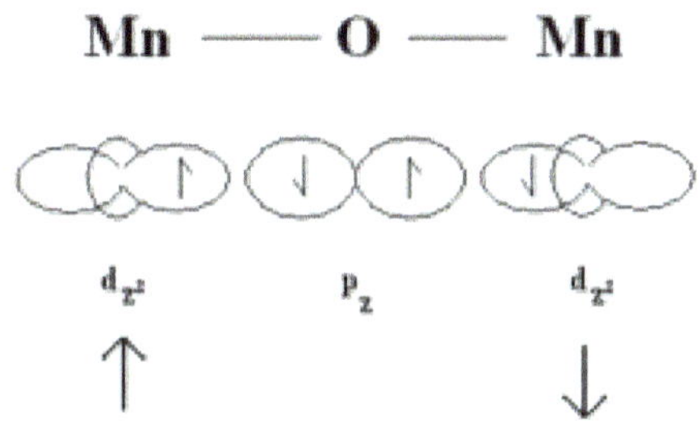

Illustration of antiferromagnetic superexchange between two transition metal cations through a shared oxygen atom.

Magnetism of ferrite spinels. Ferrite spinels are of technological interest because of their magnetic ordering, which can be ferrimagnetic or antiferromagnetic depending on the structure (normal or inverse) and the nature of the metal ions. Fe_3O_4, $CoFe_2O_4$, and $NiFe_2O_4$ are all inverse spinels and are ferrimagnets. The latter two compounds are used in magnetic recording media and as deflection magnets, respectively.

In order to understand the magnetism of ferrites, we need to think about how the unpaired spins of metal ions are coupled in oxides. If an oxide ion is shared by two metal ions, it can mediate the coupling of spins by superexchange as shown at the right. The coupling can be antiferromagnetic, as shown, or ferromagnetic, depending on the orbital filling and the symmetry of the orbitals involved. The **Goodenough-Kanamori rules** predict the local magnetic ordering (ferromagnetic vs. antiferromagnetic) that results from superexchange coupling of the electron spins of transition metal ions. For ferrites, the strongest coupling is between ions on neighboring **tetrahedral and octahedral sites**, and the ordering of spins between these two sites is reliably **antiferromagnetic**.

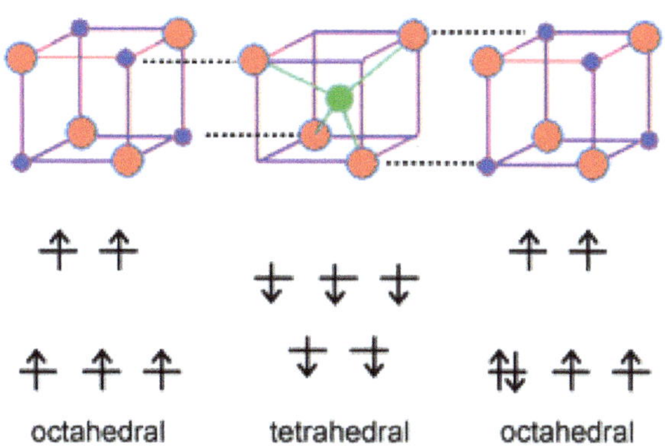

Because all the tetrahedral and octahedral sites in a spinel or inverse spinel crystal are coupled together identically, it works out that ions on the tetrahedral sites will all have one orientation (e.g., spin down) and ions on all the octahedral sites will have the opposite orientation (e.g., spin up). If the number of spins on the two sites is the same, then the solid will be antiferromagnetic. However, if the **number of spins is unequal** (as in the case of Fe_3O_4, $CoFe_2O_4$, and $NiFe_2O_4$) then the solid will be **ferrimagnetic**.

This is illustrated at the left for Fe_3O_4. The spins on the Fe^{3+} sites cancel, because half of them are up and half are down. However, the four unpaired electrons on the Fe^{2+} ions are all aligned the same way in the crystal, so the compound is ferrimagnetic.

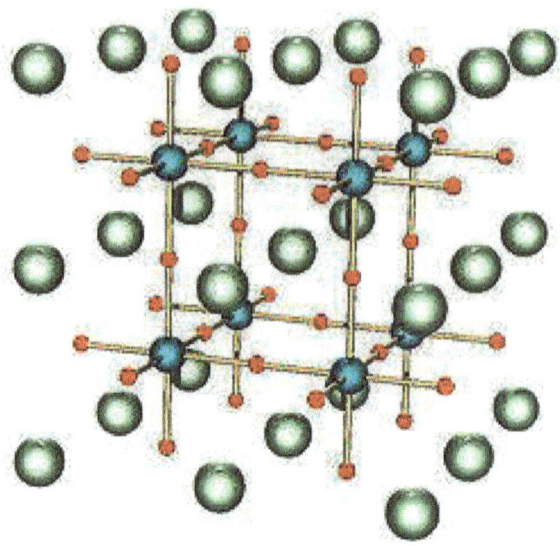

ABX_3 perovskite structure. A, B, and X are white, blue, and red, respectively.

Perovskites are **ternary oxides** of general formula **ABO_3**. More generally, the perovskite formula is ABX_3, where the anion X can be **O, N, or halogen**. The A ions are typically large ions such as Sr^{2+}, Ba^{2+}, Rb^+, or a lanthanide 3+ ion, and the B ions are smaller transition metal ions such as Ti^{4+}, Nb^{5+}, Ru^{4+}, etc. The mineral after which the structure is named has the formula $CaTiO_3$.

The perovskite structure has simple cubic symmetry, but is related to the fcc lattice in the sense that the A site cations and the three O atoms comprise a fcc lattice. The B-site cations fill 1/4 of the octahedral holes and are surrounded by six oxide anions.

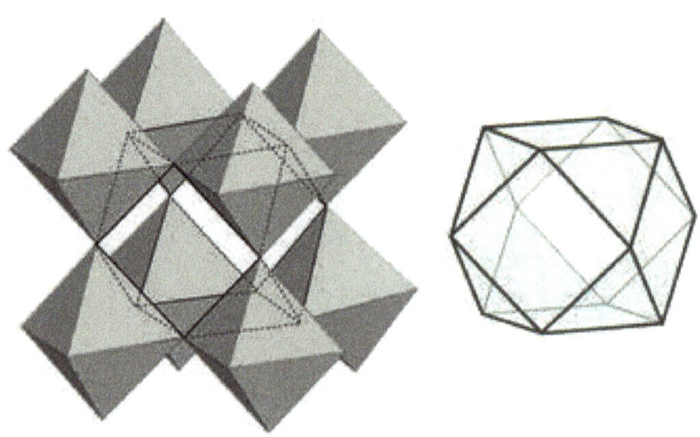

Polyhedral representation of the ReO_3 structure showing the large cuboctahedral cavity that is surrounded by 12 oxygen atoms.

The coordination of the A ions in perovsite and the arrangement of BO_6 octahedra is best understood by looking at the ReO_3 structure, which is the same structure but with the A-site cations removed. In the polyhedral representation of the structure shown at the right, it can be seen that the octahedra share all their vertices but do not share any octahedral edges. This makes the ReO_3 and perovskite structures flexible, like three-dimensional wine racks, in that the octahedra can **rotate and tilt cooperatively**. Eight such octahedra surround a large **cuboctahedral cavity**, which is the site of the A ions in the perovskite structure. Cations in these sites are coordinated by 12 oxide ions, as expected from the relationship between the perovskite and fcc lattices.

Because the A-site is empty in the ReO_3 structure, compounds with that structure can be reversibly **intercalated** by small ions such as **Li^+ or H^+**, which then occupy sites in the cuboctahedral cavity. For example, smart windows that darken in bright sunlight contain the **electrochromic** material WO_3, which has the ReO_3 structure. In the sunlight, a photovoltaic cell drives the reductive intercalation of WO_3 according to the reaction:

$$xH^+ + xe^- + WO_3 \leftrightarrows H_xWO_3$$

WO_3 is a light yellow compound containing d^0 W(VI). In contrast, H_xWO_3, which is mixed-valent W(V)-W(VI) = d^1-d^0, has a deep blue color. Such coloration is typical of mixed-valence transition metal complexes because their d-electrons can be excited to delocalized conduction band levels by red light. Because the electrochemical intercalation-deintercalation process is powered by a solar cell, the tint of the windows can adjust automatically to the level of sunlight.

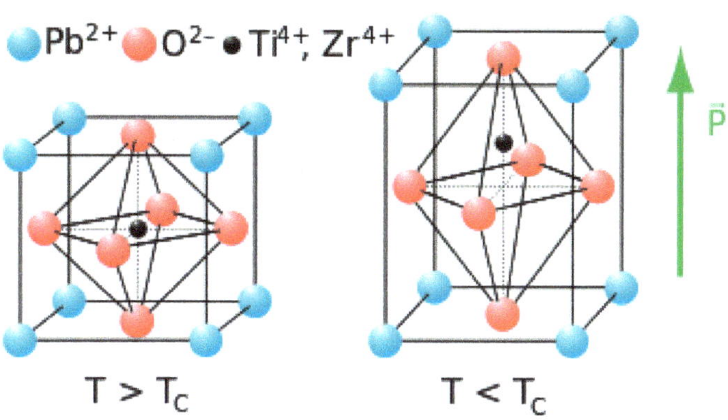

Tetragonal distortion of the perovskite unit cell in the ferroelectric oxide PZT, $PbTi_xZr_{1-x}O_3$.

Ferroelectric perovskites. The flexibility of the network of corner-sharing BO_6 octahedra is also very important in ferroelectric oxides that have the perovskite structure. In some perovsites with small B-site cations, such as Ti^{4+} and Nb^{5+}, the cation is too small to fit symmetrically in the BO_6 octahedron. The octahedron distorts, allowing the cation to move off-center. These distortions can be **tetragonal** (as in the example shown at the right), **rhombohedral**, or **orthorhombic**, depending on whether the cation moves towards a vertex, face, or edge of the BO_6 octahedron. Moving the cation off-center in the octahedron creates an **electric dipole**. In ferroelectrics, these dipoles align in neighboring unit cells through cooperative rotation and tilting of octahedra. The crystal thus acquires a net electrical polarization.

Ferroelectricity behaves analogously to **ferromagnetism**, except that the polarization is electrical rather than magnetic. In both cases, there is a **critical temperature (T_c)** above which the **spontaneous polarization** of the crystal disappears. Below T_c, the electric polarization of a ferroelectric can be switched with a coercive field, and hysteresis loop of polarization vs. field resembles that of a ferromagnet. Above T_c, the crystal is **paraelectric** and has a high dielectric permittivity.

Ferroelectric and paraelectric oxides (along with piezoelectrics and pyroelectrics) have a wide variety of applications as switches, actuators, transducers, and dielectrics for capacitors. **Ferroelectric capacitors** are important in memory devices (FRAM) and in the tuning circuits of cellular telephones. **Multiferroics**, which are materials that are simultaneously ferroelectric and ferromagnetic, are rare and are being now intensively researched because of their potential applications in electrically adressable magnetic memory.

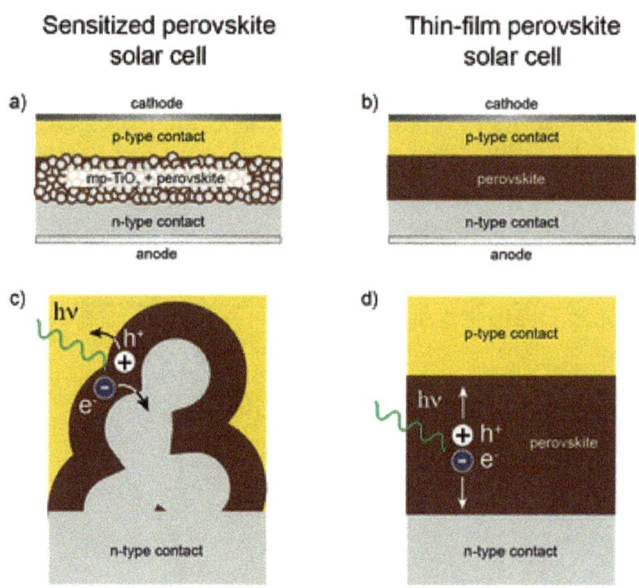

a) Solar cell architecture in which a lead halide perovskite absorber coats a layer of nanocrystalline anatase TiO_2. b) Thin-film solar cell, with a layer of lead halide perovskite sandwiched between two selective contacts. c) Charge generation and extraction in the sensitized architecture and d) in the thin-film architecture.

Halide perovskites (ABX_3, X = Cl, Br, I) can be made by combining salts of monovalent A ions ($A^+ = Cs^+$, NH_4^+, RNH_3^+) and divalent metal salts such as $PbCl_2$ or PbI_2. These compounds have sparked recent interest as light absorbers for **thin film solar cells** that produce electricity from sunlight. Lead and tin halide perovskites can be grown as thin films from solution precursors or by thermal evaporation at relatively low temperatures. In some lead halide perovskites, the mobility of electrons and holes is very high, comparable to that of more expensive III-V semiconductors such as GaAs, which must be grown as very pure single crystals at high temperatures for use in solar cells. Because of their high carrier mobility, some lead halide perovskites are also electroluminescent and are of interest as inexpensive materials for light-emitting diodes (LEDs).

Tin and lead halide perovskites were first studied in the 1990s as materials for thin film electronics,[4] and more recently as light absorbers in dye-sensitized solar cells. Soon after the results on dye-sensitized perovskite cells were reported, it was discovered that halide perovskites could also be used in thin film solid state solar cells. The structures of these solar cells are shown schematically at the right. The highest reported solar power conversion efficiencies of perovskite solar cells have jumped from 3.8% in 2009 [5] to 10.2% in 2012[6] and a certified 20.1% in 2014.[7]. The highest performing cells to date contain divalent lead in the perovskite B cation site and a mixture of methylammonium and formamidinium ions in the perovskite A cation site.

Despite their very impressive efficiency, perovskite solar cells are stable for relatively short periods of time and are sensitive to air and moisture. Current research is focused on understanding the degradation mechanisms of these solar cells and improving their stability under operating conditions.

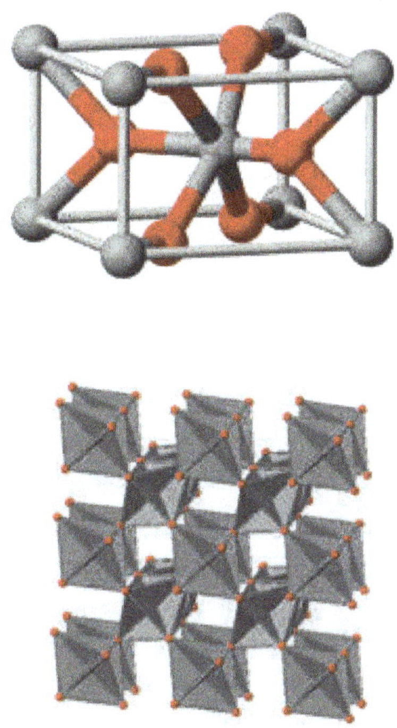

View down the tetragonal c-axis of the rutile lattice, showing edge-sharing MO_6 octahedra.

The **rutile structure** is an important MX_2 (X = O, F) structure. It is a 6:3 structure, in which the cations are octahedrally coordinated by anions, and as such is intermediate in polarity between the CaF_2 (8:4) and SiO_2(4:2) structures. The mineral rutile is one of the polymorphs of TiO_2, the others (anatase and brookite) also being 6:3 structures.

The rutile structure can be described as a distorted version of the NiAs structure with half the cations removed. Recall that compounds with the NiAs structure were typically metallic because the metal ions are eclipsed along the stacking axis and thus are in relatively close contact. In rutile, the MO_6 **octahedra share edges** along the tetragonal c-axis, and so some rutile oxides, such as **NbO_2, RuO_2 and IrO_2,** are also **metallic** because of **d-orbital overlap** along that axis. These compounds are important as electrolyzer catalysts and catalyst supports because they combine high catalytic activity with good electronic conductivity.

Rutile TiO_2, because of its high refractive index, is the base pigment for white paint. It is a wide bandgap semiconductor that has also been extensively researched as an electrode for water splitting solar cells and as a photocatalyst (primarily as the anatase polymorph) for degradation of pollutants in air and water. Self-cleaning glass exploits the photocatalytic properties of a thin film of TiO_2 to remove oily substances from the glass surface and improve the wetting properties of the glass.

REFERENCES

1. K. Mizushima, P.C. Jones, P.J. Wiseman, J.B. Goodenough (1980). "Li_xCoO_2 (0<x<l): A New Cathode Material for Batteries of High Energy Density". *Materials Research Bulletin*15: 783–789. doi:10.1016/0025-5408(80)90012-4.
2. Jump up↑ Yang Shao-Horn, Laurence Croguennec, Claude Delmas, E. Chris Nelson and Michael A. O'Keefe (July 2003). "Atomic resolution of lithium ions in $LiCoO_2$". *Nature Materials* 2 (7): 464–467. doi:10.1038/nmat922. PMID 12806387.
3. Jump up↑ M. Stanley Whittingham "Lithium Batteries and Cathode Materials" Chem. Rev., 2004, vol. 104, pp. 4271–4302. DOI: 10.1021/cr020731c
4. Jump up↑ Kagan, Cherie R.; Mitzi, David B.; Dimitrakopoulos, C. D. (1999). "Organic-inorganic hybrid materials as semiconducting channels in thin-film field-effect transistors". *Science* 286: 945-947. doi:10.1126/science.286.5441.945.
5. Kojima, Akihiro; Teshima, Kenjiro; Shirai, Yasuo; Miyasaka, Tsutomu (6 May 2009). "Organometal Halide Perovskites as Visible-Light Sensitizers for Photovoltaic Cells". *Journal of the American Chemical Society* 131 (17): 6050–6051. doi:10.1021/ja809598r. PMID 19366264.
6. Chung, In; Lee, Byunghong; He, Jiaking; Chang, Robert P. H.; Kanatzidis, Mercouri G. (24 May 2012). "All-solid-state dye-sensitized solar cells with high efficiency". *Nature* 485: 486-489. doi:10.1038/nature11067.
7. IEEE Spectrum, Perovskite Solar Cell Bests Bugbears, Reaches Record Efficiency, 7 January 2015
8. W. Zhang et al. (2013), "Unexpected Stable Stoichiometries of Sodium Chlorides." *Science*, vol. 342, no. 6165, pp. 1502-1505. doi: 10.1126/science.1244989

CHAPTER 10

Redox Stability and Redox Reactions

Chapter Contents

- **Introduction**
- **Balancing Redox Reactions**
- **Latimer and Frost Diagrams**
- **Redox Reactions with Coupled Equilibria**
- **Pourbaix Diagrams**

Redox potential, also known as standard electrode potential, is a measure of how easily a substance loses or gains electrons in a reduction-oxidation — or "redox" — reaction, a chemical reaction where one reactant is reduced and the other oxidized. Electrons move from the reducing agent to the oxidizing agent. This can be expressed as an electrical potential between the two and measured in volts. A negative value indicates a reducing agent, with a tendency to lose electrons, and a positive value indicates an oxidizing agent, with a tendency to gain electrons. The redox potential is sometimes denoted by E°.

INTRODUCTION

Oxidation states of vanadium in acidic solution. From left to right the oxidation state goes from +5 to +2.

In redox reactions, one element or compound is reduced (gains electrons) and another is oxidized (loses electrons). In terms of everyday life, redox reactions occur all of the time around us. For example, the metabolism of sugars to CO_2, which stores energy in the form of ATP, is a redox reaction. Another example of redox is fire or combustion, such as in a car engine. In a car engine, hydrocarbons in the fuel are oxidized to carbon dioxide and water, while oxygen is reduced to water. Corrosion (i.e. the formation of rust on iron) is a redox reaction involving oxidation of a metal.

Oxidation-reduction reactions are important to understanding inorganic chemistry for several reasons:

- Transition metals can have multiple oxidation states
- Main group elements (N, halogens, O, S...) also have multiple oxidation states and important redox chemistry
- Many inorganic compounds catalyze redox reactions (which are especially useful in industrial and biological applications)
- Energy conversion technologies (solar water splitting, batteries, fuel cells) rely on inorganic redox reactions and catalysis
- Electrochemistry provides a way to measure equilibrium constants for dissolution/precipitation, complexation, and other reactions.
- Reaction mechanisms in organometallic chemistry (oxidative addition, reductive elimination) involve changes in the oxidation states of metals.

Not all oxidizers and reducers are created equal. The electrochemical series ranks substances according to their oxidizing and reducing power, i.e., their standard electrode potential. Strong oxidizing agents are typically compounds with elements in high oxidation states or with high electronegativity, which gain electrons in the redox reaction. Examples of strong oxidizers include hydrogen peroxide, permanganate, and osmium tetroxide. Reducing agents are typically electropositive elements such as hydrogen, lithium, sodium, iron, and aluminum, which lose electrons in redox reactions. Hydrides (compounds that contain hydrogen in the formal -1 oxidation state), such as sodium hydride, sodium borohydride and lithium aluminum hydride, are often used as reducing agents in organic and organometallic reactions.

BALANCING REDOX REACTIONS

In studying redox chemistry, it is important to begin by learning to balance electrochemical reactions. Simple redox reactions (for example, $H_2 + I_2 \rightarrow 2\ HI$) can be balanced by inspection, but for more complex reactions it is helpful to have a foolproof, systematic method. The **ion-electron method** allows one to balance redox reactions regardless of their complexity. We illustrate this method with two examples.

Example 1: $\mathbf{I^-}$ is oxidized to $\mathbf{IO_3^-}$ by $\mathbf{MnO_4^-}$, which is reduced to $\mathbf{Mn^{2+}}$.

How can this reaction be balanced? In the ion-electron method we follow a series of four steps:

Step 1A: Write out the (unbalanced) reaction and identify the elements that are undergoing redox.

$\mathbf{MnO_4^- + I^- \rightarrow IO_3^- + Mn^{2+}}$ (The elements undergoing redox are **Mn** and **I**)

Step 1B: Separate the reaction into two **half reactions**, balancing the element undergoing redox in each.

$\mathbf{MnO_4^- \rightarrow Mn^{2+}}$

$\mathbf{I^- \rightarrow IO_3^-}$

Step 2A: Balance the **oxygen** atoms by adding water to one side of each half reaction.

$MnO_4^- \rightarrow Mn^{2+} + 4H_2O$

$3H_2O + I^- \rightarrow IO_3^-$

Step 2B: Balance the **hydrogen** atoms by adding H^+ ions.

$8H^+ + MnO_4^- \rightarrow Mn^{2+} + 4H_2O$

The left side has a net charge of **+7** and the right side has a net charge of **+2**

$3H_2O + I^- \rightarrow IO_3^- + 6H^+$

The left side has a net charge of **-1** and the right side has a net charge of **+5**

Step 2C: Balance the overall charge by adding **electrons**

$8H^+ + 5e^- + MnO_4^- \rightarrow Mn^{2+} + 4H_2O$

The left side has a charge of **+2** while the right side has a charge of **+2**. They are balanced.

$3H_2O + I^- \rightarrow IO_3^- + 6H^+ + 6e^-$

The left side has a charge of **-1** while the right side has a charge of **-1**. They are balanced.***Note:*** We did not need to explicitly determine the oxidation states of Mn or I to arrive at the correct number of electrons in each half reaction. **Step 3**: Combine the half reactions so that there are equal numbers of electrons on the left and right sides

$6\,(8H^+ + 5e^- + MnO_4^- \rightarrow Mn^{2+} + 4H_2O)$

$5\,(3H_2O + I^- \rightarrow IO_3^- + 6H^+ + 6e^-)$

$48H^+ + 30e^- + 15H_2O + 6MnO_4^- + 5I^- \rightarrow 5IO_3^- + 6Mn^{2+} + 24H_2O + 30e^- + 30H^+$Cancel the H^+, **electrons**, and **water**:$48H^+ + 30e^- + 15H_2O + 6MnO_4^- + 5I^- \rightarrow 5IO_3^- + 6Mn^{2+} + 24H_2O + 30e^- + 30H^+$

The overall balanced reaction is therefore:$\mathbf{18H^+ + 6MnO_4^- + 5I^- \rightarrow 5IO_3^- + 6Mn^{2+} + 9H_2O}$

Check your work by making sure that all **elements** and **charges** are balanced.

Step 4: If the reaction occurs under basic conditions, we add OH^- to each side to cancel $H^+$$18H^+ + 18OH^- + 6MnO_4^- + 5I^- \rightarrow 5IO_3^- + 6Mn^{2+} + 9H_2O + 18OH^-$

The $18H^+ + 18OH^-$ will become $18H_2O$ so the overall balanced reaction is:

$9H_2O + 6MnO_4^- + 5I^- \rightarrow 5IO_3^- + 6Mn^{2+} + 18OH^-$

Again, it is a good idea to check and make sure that all of the elements are balanced, and that the charge is the same on both sides. If this is not the case, you need to find the error in one of the earlier steps.

Example 2: Redox reaction of $\mathbf{S_2O_3^{2-}}$ and $\mathbf{H_2O_2 S_2O_3^{2-} + H_2O_2 \rightarrow S_4O_6^{2-} + H_2O}$

Which elements are undergoing redox? **S** and **O**<u>**Step 1**</u>: Write out half reactions, balancing the element undergoing redox

$\mathbf{2S_2O_3^{2-} \rightarrow S_4O_6^{2-}}$

$\mathbf{H_2O_2 \rightarrow 2H_2O}$

Step 2A: Balance oxygen (already balanced)

Step 2B: Balance hydrogen:

$\mathbf{2S_2O_3^{2-} \rightarrow S_4O_6^{2-}}$

$\mathbf{H_2O_2 + 2H^+ \rightarrow 2H_2O}$

Step 2C: Balance charge by adding electrons:

$\mathbf{2S_2O_3^{2-} \rightarrow S_4O_6^{2-} + 2e^-}$

$\mathbf{H_2O_2 + 2H^+ + 2e^- \rightarrow 2H_2O}$

Step 3: Combine the half reactions so that there are equal numbers of electrons on the left and right sides (already equal)

Overall balanced reaction:

$2S_2O_3^{2-} + H_2O_2 + 2H^+ \rightarrow S_4O_6^{2-} + 2H_2O$

Note that again, we did not need to know the formal oxidation states of S or O in the reactants and products in order to balance the reaction. In this case, assigning the oxidation states would be rather complex, because $S_2O_3^{2-}$ and $S_4O_6^{2-}$ both contain sulfur in more than one oxidation state.

Electrochemical Potentials

In electrochemical cells, or in redox reactions that happen in solution, the thermodynamic driving force can be measured as the **cell potential**. Chemical reactions are spontaneous in the direction of -ΔG, which is also the direction in which the cell potential (defined as $E_{anode} - E_{cathode}$) is positive. A cell operating in the **spontaneous** direction (for example, a battery that is discharging) is called a **galvanic cell**. A cell that is being driven in the **non-spontaneous** direction is called an **electrolytic cell**. For example, let us consider the reaction of hydrogen and oxygen to make water:

$2H_2(g) + O_2(g) = 2\ H_2O(l)$

Thermodynamically, this reaction is spontaneous in the direction shown and has an overall standard free energy change (ΔG°) of -237 kJ per mole of water produced.

When this reaction occurs electrochemically in the spontaneous direction (e.g., in a hydrogen-air fuel cell), the two half cell reactions that occur are:

Anode: $H_2(g) \rightarrow 2H^+(aq) + 2e^-$

Cathode: $O_2(g) + 4H^+(aq) + 4e^- \rightarrow 2\ H_2O(l)$

Here the anode is the positive electrode and the cathode is the negative electrode; under conditions of very low current density (where there are minimal resistive losses and kinetic overpotentials), the potential difference we would measure between the two electrodes would be 1.229 V.

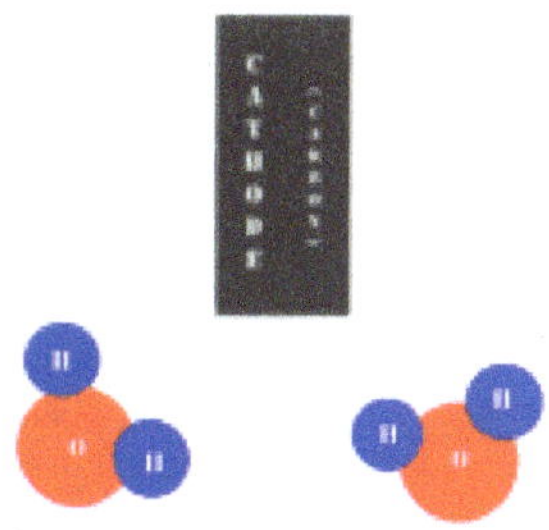

In an **electrolytic cell**, this reaction is run in reverse. That is, we put in electrical energy to split water into hydrogen and oxygen molecules. In this case, the half reactions (and their standard potentials) reverse. $O_2(g)$ bubbles form at the anode and $H_2(g)$ is formed at the cathode. Now the **anode** is the **positive** electrode and the **cathode** is **negative**. Electrons are extracted from the substance at the anode (water) and pumped into the solution at the cathode to make hydrogen. An animation of the cathode half reaction is shown at the right.

In both galvanic and electrolytic cells, **oxidation** occurs at the **anode** and **reduction** occurs at the **cathode**.

Half Reaction	Standard Potential (V)
$F_2 + 2e^- \rightleftharpoons 2F^-$	+2.87
$Pb^{4+} + 2e^- \rightleftharpoons Pb^{2+}$	+1.67
$Cl_2 + 2e^- \rightleftharpoons 2Cl^-$	+1.36
$O_2 + 4H^+ + 4e^- \rightleftharpoons 2H_2O$	+1.23
$Ag^+ + 1e^- \rightleftharpoons Ag$	+0.80
$Fe^{3+} + 1e^- \rightleftharpoons Fe^{2+}$	+0.77
$Cu^{2+} + 2e^- \rightleftharpoons Cu$	+0.34
$2H^+ + 2e^- \rightleftharpoons H_2$	0.00
$Pb^{2+} + 2e^- \rightleftharpoons Pb$	-0.13
$Fe^{2+} + 2e^- \rightleftharpoons Fe$	-0.44
$Zn^{2+} + 2e^- \rightleftharpoons Zn$	-0.76
$Al^{3+} + 3e^- \rightleftharpoons Al$	-1.66
$Mg^{2+} + 2e^- \rightleftharpoons Mg$	-2.36
$Li^+ + 1e^- \rightleftharpoons Li$	-3.05

stronger oxidizing agent

stronger reducing agent

Half-cell potentials. As noted above, the equilibrium voltage of an electrochemical cell is proportional to the free energy change of the reaction. Because electrochemical reactions can be broken up into two half-reactions, it follows that the potentials of half reactions (like free energies) can be added and subtracted to give an overall value for the reaction. If we take the **standard hydrogen electrode** as our reference, i.e., if we assign it a value of zero volts, we can measure all the other half cells against it and thus obtain the voltage of each one. This allows us to rank redox couples according to their **standard reduction potentials** (or more simply their **standard potentials**), as shown in the table on the right.

Note that when we construct an electrochemical cell and calculate the voltage, we simply take the difference between the half cell potentials and do not worry about the number of electrons in the reaction. For example, for the displacement reaction in which silver ions are reduced by copper metal, the reaction is:

$2Ag^{+}(aq) + Cu(s) = 2Ag(s) + Cu^{2+}(aq)$

The two half-cell reactions are:

$Ag^{+}(aq) + e^{-} = Ag(s)$ +0.80V

$Cu^{2+}(aq) + 2e^{-} = Cu(s)$ +0.34V

and the standard potential E° = +0.80 - 0.34 V = +0.46V

The reason we don›t need to multiply the Ag potential by 2 is that E° is a measure of the free energy change **per electron**. Dividing the free energy change by the number of electrons (see below) makes E° an **intensive property** (like pressure, temperature, etc.).

Relationship between E and ΔG. For systems that are in equilibrium, $\Delta G° = -nFE°_{cell}$, where n is number of moles of electrons per mole of products and F is the Faraday constant, ~96485 C/mol. Here the ° symbol indicates that the substances involved in the reaction are in their standard states. For example, for the water electrolysis reaction, the standard states would be pure liquid water, H^{+} at 1M concentration (or more precisely, at unit activity), and O_2 and $H_2(g)$ at 1 atmosphere pressure.

More generally (at any concentration or pressure), ΔG = -nFE , where

$$E = E^{o} - \frac{RT}{nF} * lnQ.$$

or at 298 K

$$E = E^{o} - \frac{0.0592}{n} * logQ$$

where Q is the concentration ratio of products over reactants, raised to the powers of their coefficients in the reaction. This equation (in either form) is called the Nernst equation. The second term in the equation, when multiplied by -nF, is RT*lnQ. This is the free energy difference between ΔG and ΔG°. We can think of this as an entropic term that takes into account the positive entropy change of dilution, or the negative entropy change of concentrating a reactant or product, relative to its standard state.

Using the Nernst Equation

Example 1:

For the half reaction $2H^+ + 2e^- = H_2$, $E°_{1/2} = 0.000$ V (by definition)

What is $E_{1/2}$ at pH 5 and $P_{H2} = 1$ atm?

pH = - log [H^+] = 5, so [H^+] = 10^{-5} M

$$E = E^o - \frac{0.0592}{2} * log\frac{P_{H2}}{(H^+)^2} = E^o - \frac{0.0592}{2} * log(10^5)^2 = E^o - \frac{0.0592}{2}(10)$$

$$= 0.000 - 0.296 = -0.296V$$

Toyota fuel cell hybrid bus. The bus runs on electrical energy obtained directly from the H_2/O_2 reaction. Individual fuel cells are connected in series to make a power train that charges a battery pack and drives an electric motor. Although the standard potential of the reaction is 1.23 V, because of kinetic overpotentials each fuel cell in the power train operates at a voltage of about 0.70 V. Despite this energy loss, the fuel cell system is still about twice as efficient as a combustion engine performing the same reaction.

Example 2:

What is the potential of a fuel cell (a galvanic H_2/O_2 cell) operating at pH 5?

Overall reaction: **2 H_2(g) + O_2(g) = 2 H_2O(l)**

In this reaction, H_2 is oxidized to H^+ and O_2 is reduced to H_2O. According to our convention, we write out and balance both half-cell reactions as reductions. For convenience, we do this in acid. (It is left as an exercise to the interested reader to try to work the problem in base)

Half cell reactions:

$$2\ H_2(g) = 4H^+ + 4e^-$$

$$O_2(g) + 4H^+(aq) + 4e^- = 2\ H_2O(l)$$

To solve this problem we need to find the difference between the H_2/ H^+ and O_2/ H_2O half cell potentials at pH 5.

$2H^+ + 2e^- \rightarrow H_2$ $\qquad E°_{1/2} = \quad 0.000\ V$

Like all standard potentials, this is written as a reduction. We need to reverse it and change the sign of E° since H_2 is being oxidized:

$H_2 \rightarrow 2H^+ + 2e^-$ $\qquad E°_{1/2} = -0.000\ V$

and add the standard potential of the substance being reduced at the cathode: $O_2 + 4H^+ + 4e^- \rightarrow 2H_2O$ $\quad E°_{1/2} = +1.229\ V$

The difference between the two standard half cell potentials is +1.229 - 0.000 = +1.229 V

$E°_{cell} = +1.229$ Volts

We now use the Nernst equation to account for the fact that H^+ is not in its standard state:

$$E_{cell} = E^o - \frac{0.0592}{4} * log\ \frac{[H^+]^4}{P_{O2}P_{H2}^2[H^+]^4} = E^o - \frac{0.0592}{4} * log(1) = E^o = +1.229V$$

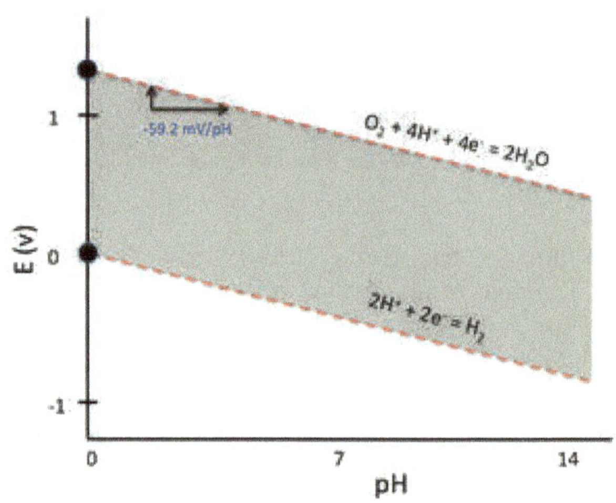

Pourbaix diagram for water.

Note that the value of E_{cell} **does not change** with pH since **both couples shift -59.2 mV/pH** according to the Nernst equation. This is the consequence of the fact that the number of electrons equals the number of protons in each of the half cell reactions. Another way to rationalize this result is to remember that the overall reaction ($2H_2 + O_2 = 2H_2O$) does not involve H^+ as a reactant or product, so ΔG and E should be independent of pH.

We can plot the shift in the H_2/H^+ and O_2/H_2O half-cell potentials with pH on a potential-pH diagram (also called a Pourbaix diagram) as shown at the left. The pH-dependent potentials of the H_2 and O_2 couples are shown as dotted lines. Notice that the potential difference between them is always 1.23 V. The dark circles represent the standard potentials.

Pourbaix diagrams are essentially electrochemical **phase diagrams**, which plot regions of thermodynamic stability for redox-active substances. As in other kinds of phase diagrams, the lines represent conditions under which two phases coexist in equilibrium. The shaded area in the water Pourbaix diagram represents the conditions of potential and pH where liquid water is stable relative to hydrogen or oxygen. Outside the shaded region, water is thermodynamically unstable and is reduced to $H_2(g)$ or oxidized to $O_2(g)$. Although these processes are spontaneous in the thermodynamic sense (for example, water is unstable in the presence of Pb^{4+}, Cl_2, Fe, Zn, or Al), they are kinetically slow and require catalysis to proceed.

LATIMER AND FROST DIAGRAMS

Manganese nodules on the sea floor contain Mn in oxidation states 2+, 3+, and 4+.

In addition to Pourbaix diagrams, there are two other kinds of redox stability diagrams known as Latimer and Frost diagrams. Each of these diagrams contains similar information, but one representation may be more useful in a given situation than the others.

Latimer and Frost diagrams help predict stability relative to higher and lower oxidation states, usually at one fixed pH. Pourbaix diagrams help understand pH-dependent equilibria, which are often coupled to solubility equilibria and corrosion (which will be talked about more later).

Latimer Diagrams:

Latimer diagrams are the oldest and most compact way to represent electrochemical equilibria for substances that have multiple oxidation states. Electrochemical potential values are written for successive redox reactions (from highest to lowest oxidation state), typically under standard conditions in either strong acid ($[H^+]$ = 1 M, pH 0) or strong base ($[OH^-]$ = 1 M, pH 14). The oxidation states of successive substances in a Latimer diagram can differ by one or more electrons. Oxidation states for the element undergoing redox are typically determined by difference; we assign the oxygen atoms an oxidation state of -2 and the hydrogen atoms an oxidation state of +1.

Example

Mn in Acid

$$\underset{7+}{MnO_4^-} \xrightarrow{+0.564} \underset{6+}{MnO_4^{2-}} \xrightarrow{+0.274} \underset{5+}{MnO_4^{3-}} \xrightarrow{+4.27} \underset{4+}{MnO_2} \xrightarrow{+0.95} \underset{3+}{Mn^{3+}} \xrightarrow{+1.51} \underset{2+}{Mn^{2+}} \xrightarrow{-1.18} \underset{0}{Mn}$$

The Latimer diagram for Mn illustrates its standard reduction potentials (in 1 M acid) in oxidation states from +7 to 0.

The Latimer diagram compresses into shorthand notation all the standard potentials for redox reactions of the element Mn. For example, the entry that connects Mn^{2+} and Mn gives the potential for the half-cell reaction:

$Mn^{2+}(aq) + 2e^- \rightarrow Mn(s)$, $E_{1/2}°$ = -1.18V

and the entry connecting Mn^{4+} and Mn^{3+} represents the reaction:

$MnO_2(s) + 4\ H^+(aq) + e^- \rightarrow Mn^{3+}(aq) + 2H_2O(l)$, :$E_{1/2}°$ = +0.95V

We can also calculate values for **multi-electron reactions** by first adding ΔG°(=-nFE°) values and then dividing by the total number of electrons

For example, for the 5-electron reduction of MnO_4^- to Mn^{2+}, we write

$$E^o = \frac{1(0.564) + 1(0.274) + 1(4.27) + 1(0.95) + 1(1.51)}{5} = +1.51V$$

and for the three-electron reduction of $MnO_4^-(aq)$ to $MnO_2(s)$,

$$E^o = \frac{1(0.564) + 1(0.274) + 1(4.27)}{3} = +1.70V$$

Remember to divide by the number of electrons involved in the oxidation number change (5 and 3 for the above equations).

Thermodynamically Stable and Unstable Oxidation States

An unstable species on a Latimer diagram will have a lower standard potential to the left than to the right.

Example:

$2MnO_4^{3-} \rightarrow MnO_2 + MnO_4^{2-}$; the MnO_4^{3-} species is unstable

E° = +4.27-0.274 = +3.997 V (spontaneous disproportionation)

Which Mn species are unstable with respect to disproportionation?

MnO_4^{3-}; 5+ → 6+,4+

Mn^{3+}; 3+ → 4+,2+

So stable species are: MnO_4^-, MnO_4^{2-}, MnO_2, Mn^{2+}, and Mn^0.

But **MnO_4^{2-}** is also unstable - why?

$MnO_4^{2-} \rightarrow MnO_2$;

$2MnO_4^{2-} \rightarrow 2MnO_4^- + MnO_2$; E° = 2.272 - 0.564 = +1.708 V

Moral: All possible disproportionation reactions must be considered in order to determine stability (this is often more convenient with a Frost diagram).

Note: Thermodynamically unstable ions can be quite stable kinetically. For example, most N-containing molecules (NO_2, NO, N_2H_4) are unstable relative to the elements (O_2, N_2, H_2), but they are still quite stable kinetically.

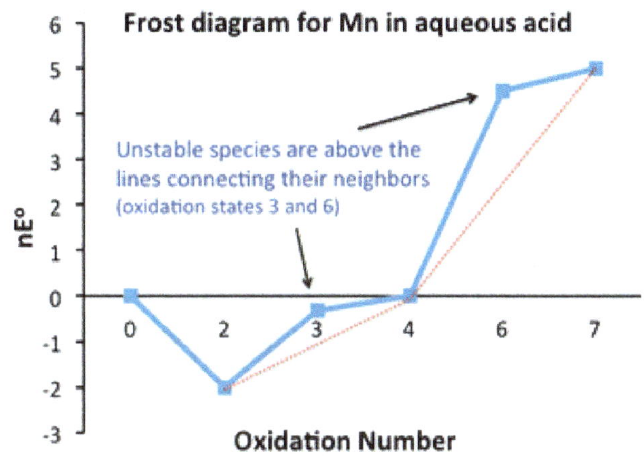

Frost diagrams: In a Frost diagram, we plot ΔG°/F (= nE°) vs. oxidation number. The zero oxidation state is assigned a nE° value of zero.

Stable and **unstable oxidation states** can be easily identified in the plot. Unstable compounds are higher on the plot than the line connecting their neighbors. Note that this is simply a graphical representation of what we did with the Latimer diagram to determine which oxidation states were stable and unstable.

The **standard potential** for any electrochemical reaction is given by the **slope** of the line connecting the two species on a Frost diagram. For example, the line connecting Mn^{3+} and MnO_2 on the Frost diagram has a slope of +0.95, the standard potential of MnO_2 reduction to Mn^{3+}. This is the number that is written above the arrow in the Latimer diagram for Mn. Multielectron potentials can be calculated easily by connecting the dots in a Frost diagram.

A Frost diagram:

Contains the same information as in a Latimer diagram, but graphically shows stability and oxidizing power.

The lowest species on the diagram are the most stable (Mn^{2+}, MnO_2)

The highest species on diagram are the strongest oxidizers (MnO_4^-)

REDOX REACTIONS WITH COUPLED EQUILIBRIA

Redox Reactions with Coupled Equilibria

Coupled equilibria (solubility, complexation, acid-base, and other reactions) change the value of E°, effectively by changing the concentrations of free metal ions. We can use the Nernst equation to calculate the value of E° from the equilibrium constant for the coupled reaction. Alternatively, we can measure the half-cell potential with and without the coupled reaction to get the value of the equilibrium constant. This is one of the best ways to measure K_{sp}, K_a, and K_d values.

As an example, we consider the complexation of Fe^{2+} and Fe^{3+} by CN^- ions:

$Fe^{2+}(aq) + 6\ CN^-(aq) \rightarrow [Fe(CN)_6]^{4-}$ (1)

$Fe^{3+}(aq) + 6\ CN^-(aq) \rightarrow [Fe(CN)_6]^{3-}$ (2)

Which oxidation state of Fe is more strongly complexed by CN^-? We can answer this question by measuring the standard half-cell potential of the $[Fe(CN)_6]^{3-/4-}$ couple and comparing it to that of the $Fe^{3+/2+}$couple:

$Fe^{3+}(aq) + e^- = Fe^{2+}(aq)$ E° = +0.77 V (3)

$[Fe(CN)_6]^{3-} + e^- = [Fe(CN)_6]^{4-}$ E° = +0.42 V (4)

Iron(III) is ***harder to reduce*** (i.e., **E° is less positive**) when it is complexed to CN^-

This implies that the equilibrium constant for complexation reaction (1) should be **smaller** than that for reaction (2). How much smaller?

We can calculate the ratio of equilibrium constants by adding and subtracting reactions:

$Fe^{3+} + 6\ CN^- \rightarrow [Fe(CN)_6]^{3-}$ $K = K_1$

$[Fe(CN)_6]^{4-} \rightarrow Fe^{2+} + 6\ CN^-$ $K = 1/K_2$

$Fe^{3+} + Fe(CN)_6^{4-} \leftrightarrow Fe^{2+} + Fe(CN)_6^{3-}$

The equilibrium constant for this reaction is the product of the two reactions we added, i.e., $K = K_1/K_2$.

But we can make the same overall reaction by combining reactions (3) and (4):

$Fe^{3+}(aq) + e^- = Fe^{2+}(aq)$ E° = +0.77 V

$[Fe(CN)_6]^{4-} = [Fe(CN)_6]^{3-} + e^-$ E° = -0.42 V

$Fe^{3+} + Fe(CN)_6^{4-} \leftrightarrow Fe^{2+} + Fe(CN)_6^{3-}$

In this case, we can calculate E° = 0.77 - 0.42 = +0.35 V

It follows from nFE° = -ΔG° = RTlnK that

$E° = RT/nF \ln(K_1/K_2)$

K_1/K_2 = exp(nFE°/RT) = exp[(1 equiv/mol)(96,500 C/equiv)(0.35 J/C)/(8.314 J/mol K)(298 K)] = exp(13.63) = **8 x 10^5**

Thus we find that $Fe(CN)_6^{3-}$ is about a million times more stable as a complex than $Fe(CN)_6^{4-}$.

Solubility Equlibria

We can use a similar procedure to measure K_{sp} values electrochemically.For example, the silver halides (AgCl, AgBr, AgI) are sparingly soluble. We can calculate the K_{sp} of AgCl by measuring the standard potential of the AgCl/Ag couple. This can be done very simply by measuring the potential of a silver wire, which is in contact with solid AgCl and 1 M Cl^-(aq), against a hydrogen reference electrode. That value is then compared to the standard potential of the Ag^+/Ag couple:

$AgCl(s) + e^- \rightarrow Ag(s) + Cl^-(aq)$ E° = +.207V

$Ag^+(aq) + e^- \rightarrow Ag(s)$ E° = +0.799 V

Subtracting the second reaction from the first one we obtain:

$AgCl(s) = Ag^+(aq) + Cl^-(aq)$ E° = +0.207 - 0.799 = -0.592 V

and again using nFE° = RTlnK, we obtain K = K_{sp} = **9.7 x 10^{-11} M^2**.

Acid mine drainage occurs when sulfide rocks such as pyrite (FeS_2) are exposed to the air and oxidized. The reaction produces aqueous Fe^{3+} and H_2SO_4. When the acidic effluent of the mine meets a stream or lake at higher pH, solid $Fe(OH)_3$ precipitates, resulting in the characteristic orange muddy color downstream of the mine. The acidic effluent is toxic to plants and animals.

Because the solubility of the silver halides is so low, this would be a very difficult number to measure by other methods, e.g., by measuring the concentration of Ag+ spectroscopically, or by gravimetry. In practice almost all K_{sp} values involving electroactive substances are measured potentiometrically.

Acid-Base Equilibria

Many electrochemical reactions involve H^+ or OH^-. For these reactions, the half-cell potentials are **pH-dependent**.

Example: Recall that the disproportionation reaction $3MnO_4^{2-}(a) \rightarrow 2MnO_4^-(aq) + MnO_2(s)$ is **spontaneous** at pH=0 ($[H^+] = 1M$), from the Latimer diagram or Frost plot.

However, when we properly balance this half reaction we see that it involves **protons as a reactant**:

$3\ MnO_4^{2-}(aq) + \mathbf{4H^+}(aq) \leftrightarrow 2MnO_4^-(aq) + MnO_2(s) + 2H_2O(l)$

By Le Châtelier's principle, it follows that **removing protons** (increasing the pH) should **stabilize** the reactant, $\mathbf{MnO_4^{2-}}$. Thus we would expect the +6 oxidation state of Mn, which is unstable in acid, to be stabilized in basic media. We will examine these proton-coupled redox equilibria more thoroughly in the context of Pourbaix diagrams below.

POURBAIX DIAGRAMS

Pourbaix Diagrams plot electrochemical stability for different redox states of an element as a function of pH. As noted above, these diagrams are essentially phase diagrams that map the conditions of potential and pH (most typically in aqueous solutions) where different redox species are stable. We saw a simple example of such a diagram in section 4.2 for H_2O. Typically, the water redox reactions are plotted as dotted lines on these more complicated diagrams for other elements.

The lines in Pourbaix diagrams represent redox and acid-base reactions, and are the parts of the diagram where two species can exist in equilibrium. For example, in the Pourbaix diagram for Fe below, the horizontal line between the Fe^{3+} and Fe^{2+}regions represents the reaction $Fe^{3+}(aq) + e- = Fe^{2+}(aq)$, which has a standard potential of +0.77 V. While we could use standard potentials for all these lines, but in practice Pourbaix diagrams are usually plotted for lower ion concentrations (often 1 mM) that are more relevant to corrosion and electrochemical experiments.

Example: Iron Pourbaix diagram

Areas in the Pourbaix diagram mark regions where a single species ($Fe^{2+}(aq)$, $Fe_3O_4(s)$, etc.) is stable. More stable species tend to occupy larger areas.

Lines mark places where two species exist in equilibrium.

- **Pure redox** reactions are **horizontal** lines - these reactions are not pH-dependent
- **Pure acid-base** reactions are **vertical** lines - these do not depend on potential
- Reactions that are **both** acid-base and redox have a slope of -0.0592 V/pH x # H^+/# e^-)

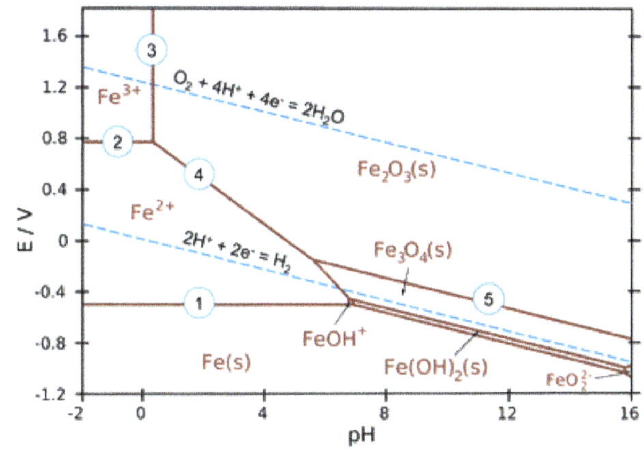

Pourbaix diagram for iron at ionic concentrations of 1.0 mM.

Examples of equilibria in the iron Pourbaix diagram (numbered on the plot):

- **1.** $Fe^{2+} + 2e^- \rightarrow Fe(s)$ (pure redox reaction - no pH dependence)
- **2.** $Fe^{3+} + e^- \rightarrow Fe^{2+}$ (pure redox reaction - no pH dependence)
- **3.** $2\ Fe^{3+} + 3\ H_2O \rightarrow Fe_2O_3(s) + 6H^+$ (pure acid-base, no redox)
- **4.** $2\ Fe^{2+} + 3\ H_2O \rightarrow Fe_2O_3(s) + 6H^+ + 2e^-$ (slope = -59.2 x 6/2 = -178 mV/pH)
- **5.** $2\ Fe_3O_4(s) + H_2O \rightarrow 3\ Fe_2O_3(s) + 2H^+ + 2e^-$ (slope = -59.2 x 2/2 = -59.2 mV/pH)

The water redox lines have special significance on a Pourbaix diagram for an element such as iron. Recall that liquid water is stable *only* in the region between the dotted lines. Below the H_2 line, water is unstable relative to hydrogen gas, and above the O_2 line, water is unstable with respect to oxygen. For active metals such as Fe, the region where the pure element is stable is typically below the H_2 line. This means that iron metal is unstable in contact with water, undergoing reactions:

$Fe(s) + 2H^+ \rightarrow Fe^{2+}(aq) + H_2$ (in acid)

$Fe(s) + 2\ H_2O \rightarrow Fe(OH)_2(s) + H_2$ (in base)

Iron (and most other metals) are also thermodynamically unstable in air-saturated water, where the potential of the solution is close to the O_2 line in the Pourbaix diagram. Here the spontaneous reactions are:

$4\ Fe(s) + 3\ O_2 + 12H^+ \rightarrow 4\ Fe^{3+} + 6\ H_2O$ (in acid)

$4\ Fe(s) + 3\ O_2 \rightarrow 2\ Fe_2O_3(s)$ (in base)

Corrosion and passivation. It certainly sounds bad for our friend Fe: unstable in water, no matter what the pH or potential. Given enough time, it will all turn into rust. But iron (and other active metals) can corrode, or can be stabilized against corrosion, depending on the conditions. Because our civilization is dependent on the use of active metals such as Fe, Al, Zn, Ti, Cr... for practically everything, it is important to understand this, and we can do so by referring to the Pourbaix diagram.

The **corrosion** of iron (and other active metals such as Al) is indeed rapid in parts of the Pourbaix diagram where the element is oxidized to a soluble, ionic product such as Fe^{3+}(aq) or Al^{3+}(aq). However, solids such as Fe_2O_3, and especially Al_2O_3, form a protective coating on the metal that greatly impedes the corrosion reaction. This phenomenon is called **passivation.** Draw a vertical line through the iron Pourbaix diagram at the pH of tap water (about 6) and you will discover something interesting: at slightly acidic pH, iron is quite unstable with respect to corrosion by the reaction:

$$Fe(s) + 2H^+ \rightarrow Fe^{2+}(aq) + H_2$$

but *only in water that contains relatively little oxygen,* i.e., in solutions where the potential is near the H_2 line. Saturating the water with air or oxygen moves the system closer to the O_2 line, where the most stable species is Fe_2O_3 and the corrosion reaction is:

$$4\ Fe(s) + 3\ O_2 \rightarrow 2\ Fe_2O_3(s)$$

This oxidation reaction is orders of magnitude slower because the oxide that is formed *passivates* the surface. Therefore iron corrodes much more slowly in oxygenated solutions.

More generally, iron (and other active metals) are passivated whenever they oxidize to produce a solid product, and corrode whenever the product is ionic and soluble. This behavior can be summed up on the color-coded Pourbaix diagram below:

In the yellow part of the diagram, an active metal such as iron can be **protected** by a second mechanism, which is to bias it so that its potential is below the oxidation potential of the metal. This **cathodic protection** strategy is most frequently carried out by connecting a more active metal such as Mg or Zn to the iron or steel object (e.g., the hull of a ship, or an underground gas pipeline) that is being protected. The active metal (which must be higher than Fe in the activity series) is also in contact with the solution and slowly corrodes, so it must eventually be replaced. In some cases a battery - the anode of which oxidizes water to oxygen in the solution - is used instead to apply a negative bias.

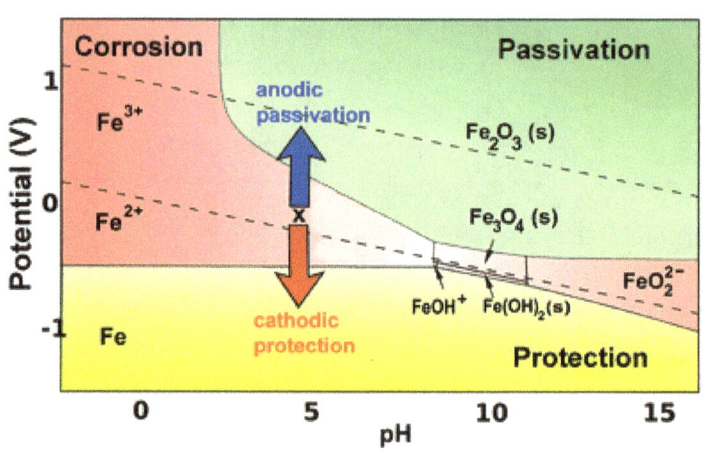

Rusty cast iron bridge and lamppost, North Ayrshire, Scotland.

Another common mode of corrosion of iron and carbon steel is **differential aeration.** In this case, part of the iron object - e.g., the base of a bridge, or the drill in an oil rig - is under water or in an anoxic environment such as mud or soil. The potential of the solution is close to the H_2 line in the Pourbaix diagram, where Fe can corrode to Fe^{2+} (aq). Another part of the iron object is in the air, or near the surface where water is well oxygenated. At that surface oxygen can be reduced to water, $O_2 + 4H^+ + 4e^- = 2\ H_2O$. The conductive iron object completes the circuit, carrying electrons from the anode (where Fe is oxidized) to the cathode (where O_2 is reduced). Corrosion by differential aeration can be rapid because soluble ions are produced, and the reaction has a driving force of over 1 V. Iron or carbon steel that is subjected to frequent weathering, such as the cast iron bridge and lamppost shown at the right, is corroded on the surface by differential aeration.

Differential aeration is involved in the formation of a rust ring around wet areas of cast iron, e.g., an iron frying pan left partially submerged in water for a day or more. Under the water, Fe is oxidized to soluble Fe^{2+}, and at the water line O_2 is reduced to H_2O. As Fe^{2+} ions diffuse towards the water surface, they encounter oxygen molecules and are oxidized to Fe^{3+}. However Fe^{3+} is insoluble at neutral pH and deposits as rust, typically just below the water line, forming the rust ring.

REFERENCES

1. Rosi, Nathaniel L.; Eckert, Juergen; Eddaoudi, Mohamed; Vodak, David T.; Kim, Jaheon; O'Keefe, Michael; Yaghi, Omar M. (2003). "Hydrogen storage in microporous metal-organic frameworks". *Science* **300** (5622): 1127–1129. doi:10.1126/science.1083440. PMID 12750515. Bibcode: 2003Sci...300.1127R.

2. "Metal–Ligand Multiple Bonds: The Chemistry of Transition Metal Complexes Containing Oxo, Nitrido, Imido, Alkylidene, or Alkylidyne Ligands W. A. Nugent and J. M. Mayer; Wiley-Interscience, New York, 1988.
3. McConville, David H.; Wolf, Jennifer R.; Schrock, Richard R. (1993). Synthesis of chiral molybdenum ROMP initiators and all-cis highly tactic poly(2,3-(R)2norbornadiene) (R = CF_3 or CO_2Me)". *J. Am. Chem. Soc.* **115** (10): 4413–4414. doi:10.1021/ja00063a090.
4. Nguyen, Sonbinh T.; Johnson, Lynda K.; Grubbs, Robert H.; Ziller, Joseph W. (1992). Ring-opening metathesis polymerization (ROMP) of norbornene by a Group VIII carbene complex in protic media. *J. Am. Chem. Soc.* **114** (10): 3974–3975. doi:10.1021/ja00036a053.
5. Rosenberg B, Vancamp L, Krigas T (1965). Inhibition of cell division in Escherichia coli by electrolysis products from a platinum electrode. *Nature* **205** (4972): 698–699. doi:10.1038/205698a0.
6. G. Wilkinson, M. Rosenblum, M. C. Whiting, R. B. Woodward (1952). "The Structure of Iron Bis-Cyclopentadienyl. *Journal of the American Chemical Society* **74** (8): 2125–2126. doi:10.1021/ja01128a527.
7. Osborn, J. A.; Jardine, F. H.; Young, J. F.; Wilkinson, G. (1966). "The Preparation and Properties of Tris(triphenylphosphine)halogenorhodium(I) and Some Reactions Thereof Including Catalytic Homogeneous Hydrogenation of Olefins and Acetylenes and Their Derivatives. *Journal of the Chemical Society A*: 1711–1732. doi:10.1039/J19660001711.
8. "Tris(triphenylphosphine)halorhodium(I) J. A. Osborn, G. Wilkinson, Inorganic Syntheses, 1967, Volume 10, p. 67. DOI 10.1002/9780470132418.ch12
9. D. A. Evans, G. C. Fu and A. H. Hoveyda (1988). Rhodium(I)-catalyzed hydroboration of olefins. The documentation of regio- and stereochemical control in cyclic and acyclic systems. *J. Am. Chem. Soc.* **110** (20): 6917–6918. doi:10.1021/ja00228a068.
10. I. Ojima, T. Kogure (1972). Selective reduction of α,β-unsaturated terpene carbonyl compounds using hydrosilane-rhodium(I) complex combinations. *Tetrahedron Lett.* **13** (49): 5035–5038. doi:10.1016/S0040-4039(01)85162-5.
11. W. S. Knowles (2003). "Asymmetric Hydrogenations (Nobel Lecture 2001)". *Advanced Synthesis and Catalysis* **345** (12): 3–13. doi:10.1002/adsc.200390028.
12. H. Jahn and E. Teller (1937). "Stability of Polyatomic Molecules in Degenerate Electronic States. I. Orbital Degeneracy". *Proceedings of the Royal Society A* **161** (905): 220–235. doi:10.1098/rspa.1937.0142. Bibcode: 1937RSPSA.161..220J.
13. Rob Janes and Elaine A. Moore (2004). *Metal-ligand bonding*. Royal Society of Chemistry. ISBN 0-85404-979-7.
14. Patrick Frank, Maurizio Benfatto, Robert K. Szilagyi, Paola D›Angelo, Stefano Della Longa, and Keith O. Hodgson The Solution Structure of $[Cu(aq)]^{2+}$ and Its Implications for Rack-Induced Bonding in Blue Copper Protein Active Sites" Inorganic Chemistry 2005, vol 44, pp 1922–1933. DOI 10.1021/ic0400639

Index

V